CABINET=MAKING

FOR AMATEURS.

CABINET=MAKING

FOR AMATEURS.

A PRACTICAL HANDBOOK ON THE
MAKING OF VARIOUS ARTICLES
OF FURNITURE.

BY VARIOUS HANDS.

EDITED BY

JOHN P. ARKWRIGHT.

𝔍llustrateð.

University Press of the Pacific
Honolulu, Hawaii

Cabinet-Making for Amateurs:
A Practical Handbook on the Making of Various
Articles of Furniture

Edited by
John P. Arkwright

ISBN: 1-4101-0231-9

Fredonia Books
Amsterdam, The Netherlands
http://www.fredoniabooks.com

PREFACE.

THE number of Amateur Carpenters is almost untold, and the vast majority of them are to a greater or lesser extent Cabinet Makers also, but there has been no handy practical book to help them in their work. This little book is an effort to remedy the omission. How far it may succeed or fail, time and the public will show; but, at any rate, the fact will remain that within its covers will be found sufficiently full instructions for anyone who is not an absolute novice at carpentry work to construct a considerable number of useful and ornamental articles. Some of them, no doubt, are not, strictly speaking, "cabinet work," but they have been considered to be sufficiently near it to be included in this volume.

J. P. A.

CONTENTS.

CABINET-MAKING FOR AMATEURS.

THE CARPENTER'S BENCH.

IN this chapter we shall give particulars of two benches—a very simple one and a complete one.

The first one, though it is, as may be surmised, more or less a makeshift, has proved quite serviceable, and very few of required to make it is such as everyone who wants to have a bench will find within his powers, and the tools required are few.

Fig. 1 shows the bench or work-table as it stands. The dimensions are as

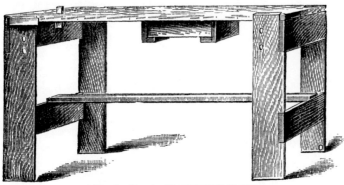

FIG. 1.—SIMPLE CARPENTER'S BENCH.

the articles of furniture described in these pages could not be made on it. The cost of material is of the smallest, and if it is not so firm and rigid as a carpenter's heavy bench, it is, at any rate, sufficiently so for ordinary purposes. The skill follows: Top, 4ft. 6in. by 1ft. 6in.; height, 2ft. 6in. These, of course, can be varied to suit special requirements, at the same time preserving the same general proportions. The material used throughout is pine (common spruce) of

1in. thickness. The exact thickness is not of much consequence, and those who contemplate doing any heavy work on the bench may as well make it of stouter stuff. To get the necessary width of top two pieces of board are glued together and properly squared off at the ends, back and front.

Four pieces for the legs, which, of course, must be of the length named for the bench height, will be required, and about 5in. square. Four similar pieces of the same length as the width of the top will also have to be got out to connect each end pair of legs.

As will be seen by reference to Fig. 1, one of these pieces at each end is immediately under the top, which is nailed to it, the remaining pieces being attached to the legs lower down. Connecting these lower pieces is another, which not only serves as a stay but as a shelf for tools, &c.

All the parts being ready, the next thing will be to fit them together, for they should only be temporarily fastened till it has been ascertained that everything is correct.

Fig. 2 shows the joints connecting the legs to the top, and those of what

FIG. 2.—LEG AND TOP JOINT.

we may call the framing-pieces connecting the back and front legs, and on which the top rests. The illustration is so clear that verbal description almost seems superfluous; but for the benefit of those who are not able to recognise the construction, a few words of explanation may be given. On the outer edge of each leg a recess is cut to admit the piece which has been alluded to as the framing. The leg, however, runs through the top, or, to express the meaning in other and perhaps more generally understood words, the leg is fitted, not under the top, but against it. The space cut in the leg, therefore, must not be the

length required to admit the frame-piece only, but must take in the thickness of the top as well. It will thus be seen that to get the length of the space, the width of the frame-piece and the thickness of the top must be added together, when, by marking off with square and gauge, this latter set to the thickness of the frame, the exact space to insure a well-fitting joint can be obtained accurately, and without the annoying necessity for wasting time by subsequent paring and chiselling.

After this explanation it will be easy to see that the legs are fitted into the top in a precisely similar manner. The space for the now somewhat narrowed top end of the leg is marked off with the square, and the thickness of the legs set out with the gauge. A couple of saw cuts are made along the lines indicated by the square, while the gauged line shows the depth to which the superfluous wood is removed with the chisel. In doing this it will simplify matters to make several saw cuts as far as the gauged line, so that small pieces can be chopped out without danger of splitting the wood through any inequality of grain. It goes almost without saying that the remaining two end pieces are let into the legs in the same manner. Their precise position is not important, but if they are about 1ft. from the ground, that is to their top edges, they will do very well. If a tool-chest or box is to be placed under the shelf, it will be as well to regulate the height accordingly.

When we have got this length, the parts may be fastened together, if the tentative fitting-up has shown that everything is right. If not, the necessary alterations should be made, though it can only be through some great neglect if there is anything seriously wrong, or any defect, such, perhaps, as levelling the legs, which cannot be managed more easily afterwards.

To fasten the pieces together either nails or screws may be used, the latter being preferable for some reasons; they, however, take more time to fit, and if good-sized nails are used and driven in on the skew, or in slanting directions, they will do very well. It does not much matter in what order the fixing is proceeded with, but perhaps the following will be the most convenient: Fasten the legs to the top, using four good nails to each, then take the top framing-pieces, nailing them to the legs in a similar manner, and fastening the top down to them with nails 3in. or so apart. The lower end pieces may then be nailed to the legs, and the shelf fastened down on top of them.

We have now got a table or bench, but it is still incomplete. For planing up the surface of boards, for instance, a stop is wanted. This, of course, must be somewhere near the left hand end of the top. All that is necessary is a piece of hard wood of from 4in. to 6in. in length, 2in. wide, and 1½in. thick. It will be understood that these sizes may be departed from to almost any extent, but those given will be found generally convenient and suitable. A hole in which the stop will fit tightly—the tighter the better—must be cut through the top with a chisel. To raise it to a convenient height for any piece of work to be "stopped" by it, it is only necessary to hammer it up from below, while if it is required out of the way at any time, a few taps from above with the same tool will do all that is required.

Another thing of equal value or even greater importance than the stop in a carpenter's bench is some arrangement by which boards may be held and supported while their edges are being planed up, as well as during sundry other details of construction. This contrivance usually takes the form of a movable block and one or two wooden screws working in a block fitted below the top, the whole being known as a bench-screw. Although one of these may be bought for a small sum, it will be more in accordance with the present purpose to describe another arrangement by which the use of screws is obviated; it is not in all respects so satisfactory as a bench-screw, but it is, nevertheless, thoroughly efficient for ordinary work, and it, moreover, possesses

FIG. 3. SIDE STOP.

the merit of being easily constructed in its entirety by the unskilled amateur. It will be recognised on Fig. 1 at the left-hand front corner of the top, while Fig. 3 shows it on an enlarged scale in plan—i.e., looking down on it from above. This clamp or side-stop is formed of a piece of hard, tough wood, and its size must chiefly depend on the uses to which it is to be put. For planing purposes—i.e., for shooting edges of boards—it need not be so long as if it is intended to hold work as in a vice. From the description, however, no one will have any difficulty in modifying sizes and proportions to suit himself, while those given will be found suitable enough for ordinary joinery purposes. Let it then be about 6in. long, 2in. wide, and not less than 1in. thick. From this cut out the wedge-shaped space as shown in Fig. 3, and fasten securely to the bench at the extreme left-hand end of the front edge of the top. The screws should be of good substance and length, entering the bench top for at least 1in.

In connection with this stop it will be necessary to have some support for the boards being worked on, and for this purpose the two front legs must be made to hold the pegs on which to rest the wood. The holes may be bored through the legs with a centre-bit, just wherever may happen to be most convenient, and a couple of loose pegs —one for each leg—made to fit. The pegs should be stout enough to bear any weight likely to be placed on them.

Now, supposing we have a piece of board, the edges of which must be planed : it is taken and laid on the pegs, these being so arranged that the edge is a little above the level of the bench-top ; one end is forced into the wedge-shaped opening in the stop, and we have nearly all the convenience of a screw.

It is obvious that a board to be supported in this way must be almost as long as the bench ; but it will frequently happen that the edges of much shorter pieces require planing. This may easily be managed by laying a piece of wood temporarily on the pegs, so that the piece of wood to be planed can be supported on it.

A bench made as described is somewhat light and easily shifted, but this little difficulty is easily got over by the use of a couple of iron angle brackets fixed at any convenient place to the legs and floor, or if those of metal should not be handy, all that is wanted can be made from two or three odd bits of wood.

Benches are commonly provided with a sunk recess or tray at the back, in which tools may be laid, and the amateur will find it handy to have some arrangement for this purpose It may easily be managed by making a long shallow box or trough, and nailing it to the back legs. Its size is unimportant; but if it is 5in. or 6in. wide by slightly less in depth it will do very well.

A drawer or sliding tray underneath the top and opening to the front will be a convenience for sundry small tools, odds and ends, &c. It should not be more than 3in. or 4in. deep, made of fairly stout wood, and nailed or screwed together, unless the maker wishes to try his hand at dovetailing.

though, naturally, attention must be paid to the joinery work, for if the joints are badly made, it cannot be expected that there will be much rigidity.

The so-called German bench is generally considered to be an improvement on the ordinary form met with in this country. Those who are acquainted with it will see that the bench shown in Fig. 4 is a modification, for though simpler than the German in its best form, it preserves its chief characteristic in having a bench-screw at the end as well as in the front.

Roughly speaking, the length of the bench may be whatever the worker thinks will suit him best, but a few hints will not be amiss. Probably the length

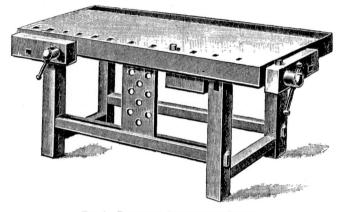

FIG. 4.—COMPLETE CARPENTER'S BENCH.

The method of fixing is seen in Fig. 1, which shows the bench-top, below which are fastened two pieces as long as, and just a trifle wider than, the drawer sides, and along their lower edges two pieces of any width on which the drawer may run. This arrangement may seem rough and ready; but, like all the other parts, it will be found thoroughly efficient, and not to be despised on account of its crudeness.

The bench which will now be described is just such a one as will suit the majority of both amateur and professional workmen, and is of a much more substantial and complete character than the somewhat make-shift one previously described. Its construction is not difficult,

that will be most generally acceptable may be considered as 5ft. or 5ft. 6in. This may be regarded as the happy medium, for 7ft. would be a maximum and 4ft. a minimum length. Irrespective of length, the width of the top may be anything between 1ft. 4in. and 2ft., exclusive of the recess at the back. As a rule there is no advantage in having a very wide top, and for a bench 5ft. long a width of 18in. will be ample. Mr. J. D. Sutcliffe, in his "Handcraft"—a work which may be regarded as a textbook of Slojd—says that the (Slojd) "bench is usually about 7ft. long, 2ft. wide, and 3ft. 3in. high." These dimensions seem unnecessarily large, but they may be a guide to some. In height 3ft.

will be found to be convenient for most people.

There is considerable choice in the kind of wood. Generally speaking, nothing could be better than beech, which is not only fairly inexpensive, but is in every way suitable and easily obtainable. If beech cannot be got, or something cheaper is wanted, pine will do very well, the principal objection to it being its comparative softness. For the legs and under-part, however, it will do as well as anything. Whatever the kind of wood it should be thoroughly well seasoned, dry stuff, and sound, or free from shakes or cracks and knots. For the top especially a nice, sound piece should be selected. It ought not to be less than 4in. thick if a really first-class bench be wanted, on which any kind of joinery work may be done; but there is no absolute necessity for having this thickness. Let it then be assumed that a suitable piece of wood for the bench-top, from 2in. to 4in. thick, has been obtained. It should be made perfectly level on the upper side at least, the edges ought also to be squared up, and the whole nicely smoothed by means of the plane. The top or main portion being to a certain extent ready, the legs and under-framing may be attended to.

The legs themselves may very suitably be 3in. square, or rather be cut from wood nominally of that thickness, for when planed down they will not measure so much. They should be of exactly the same length, and be properly squared off at the ends. They may be mortised and tenoned into the top, but it may be more convenient, in case the bench has to be moved at any time, to frame them up, so that the top can be screwed and unscrewed to them.

The pieces connecting each pair of legs at the ends may either be of the same

FIG. 5.—MORTICE-AND-TENON-JOINT.
(Showing A, Mortice in End of Leg; B, Tenon in End of Frame)

substance as the legs themselves, or any measurement between that and 1in. One

inch square, however, would not do, for in order to give the necessary stability these pieces should be not less than 3in. wide. They should be mortised and tenoned to the legs, or perhaps dove-tailed would be better, though slightly more difficult for a novice. Figs. 5 and 6

FIG. 6.—DOVETAIL JOINT.
(Showing A, Socket in End of Leg , B, Dovetail on End of Frame)

show these two joints respectively. If preferred, from the comparative ease with which it may be done, these pieces can be simply screwed to the legs, as shown in Fig. 7. This will not make such a neat, workman-like job, but it will be quite as strong as the other, or indeed stronger, unless the joints are well made and close-fitting. Near the foot end of the legs, say 3in. or 4in. from

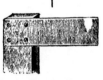

FIG. 7.—SCREW FRAMING.

the ground, exactly similar pieces should be fitted and securely fastened. At this stage then there will be two end trestles ready as shown in Fig. 8. If the bench is only a narrow one, it will be as well to make the trestles as wide as possible, not merely under the main portion of the top, but coming under the tray at the back, that extra stability may be gained.

The bench-screws can be got at a very moderate price at most tool-shops, so that it is not worth while to attempt to make them, unless the necessary box and tap for cutting the threads are available. They can, in fact, be bought complete for as little as the amateur would probably have to pay for the wood to make them of. To use the bench-screw two "chops," as they are called, are necessary. They are the flat pieces between which the wood to be worked on is held. One of them has a thread cut in it to correspond with that on the screw; the other has only a plain hole cut through it within which the screw

FIG. 8.—END TRESTLE.

runs loosely. Very likely the threaded block will be the only one obtainable with the screw, but the other one can easily be made. The exact position in which the threaded block is placed is a matter worthy of some consideration, for the lower it is placed the wider a piece of wood can be gripped between the bench and the loose block above the screw. We may suppose that the place selected is immediately under the top, or even higher still, let into the edge of the top, and the two methods may be briefly described.

If the block is to be fixed under the top, it may either be screwed from above, and well secured by supplementary blocks fastened on behind, or fixed in the angle at the juncture of the leg and the top, in which case probably no additional support will be necessary.

Whether the block can be let into the edge of the top or not, will to a certain extent depend both on the thickness of the top and on the size of the block, for it must be understood that the top of the latter must be flush with the surface of the bench. If, therefore, the block is so small that this cannot be managed without interfering with the screw, a groove must be cut in the lower surface of the top for this to work in. With a

thin top this groove will probably be unnecessary. The best way to fix the block into the top will be by means of a dovetail joint, as shown in Fig. 9. In this case, too, as will be seen from the illustration, advantage may be taken of the leg to obtain additional support.

The block at the end will, of course, be fastened in a similar way, and to make room for the screw a hole may have to be cut in the framing. For this reason it was recommended that the framing-pieces should be of a good width. If the legs be set well in from the end, there may be no occasion to cut into the framing; but on these and other similar details the best guide will be the maker's common sense, for it is obviously impossible, without knowing exactly the means at his command, to do more than indicate the various matters which must be attended to. Care should be taken that the screw works parallel with the upper surface of the bench, or as nearly so as practicable.

The next thing to be attended to may as well be the loose chop, or block, between which and the bench any piece

FIG. 9.—SCREW BLOCK FIXED TO BENCH BY DOVETAIL JOINT.

of wood to be worked is held. Fig. 4 shows clearly enough the requisite proportions of this. The wood should be hard and sound—*i.e.*, free from cracks, or shakes as they are called. It must be thick enough not to give at all under the strain of the screw, and it should hardly be less than 2in. In width it should not be less than the thickness of the bench-top if this is a thick one, and 1in. to 2in. more will be rather an advantage than otherwise by saving labour. If the bench-top is only a thin one, by which is meant anything less than 3in., the block should be at least 2in. wider. The reason for this width is to allow of the guide, which is necessary with a single screw, being fitted to it. The guide is dispensed with when a double screw-block is used, but is absolutely necessary with such a one as we are supposing; otherwise it would be impossible to keep the loose chop in its proper position. In length, the block may be anything in reason. To the left of the screw it must be long enough to permit the guide to be fastened at a sufficient distance—say 6in.—from the screw. On the right it must be sufficiently long to enable it to hold whatever may be required between it and the bench-top. Probably from 4in. to 6in. will be enough. Excessive length will only be an inconvenience.

There are various ways in which the guide may be fitted, and it may be either of wood or iron, as may suit the maker best, and be round or square. If the latter, a piece of an ordinary broom-handle will do as well as anything. Of whatever kind it is, it should not be too short, and, as a general direction, the most convenient length may be stated as a little less than the width of the bench-top. The principal thing is to have it long enough to run freely in the bearings which will be prepared for it below the bench-top. Should a square guide be preferred, a piece of tough wood, 1in. by 1in., will do; but it will not be objectionable if it is more. A piece of iron rod, ⅞in. thick, was once used successfully on a bench the writer made for himself. It must be securely fastened to the loose block, either by being let into a hole bored through this or underneath it. If the block is only the width of the thickness of the top it will be necessary to put the guide in the latter position, unless the alternative of cutting a groove in the lower surface of the top be not objected to. The fitting of the guide must, in fact, be considered in conjunction with the bearings and their exact position. When fastening the guide into the block, care should be taken that it will run parallel with the screw.

Like the guide, the bearings allow of considerable choice in their details. Supposing the bearing is to be in the groove under the bench-top previously mentioned then it may be covered at the bottom by having a piece of wood nailed over it. If the groove is not deep enough to let the whole bulk of the guide lie within it, it is obviously not a very difficult matter to cut a corresponding groove in the bottom piece of wood. Should it not be necessary or advisable to cut a groove in the top, the same object can be attained by making, as it were, a long, narrow box open at the ends and on top, with the internal dimensions just large enough for the guide to move easily lengthwise. There will be no difficulty in fastening such a bearing in its proper place underneath the top. If it is wanted a little lower down than immediately under the top, all that is required is a piece of wood of suitable thickness intervening. It is not necessary for such a casing to extend the whole length of the guide. If there is sufficient bearing to keep the guide in its proper place, nothing more is wanted. A continuous bearing is not essential, and the desired object may even be attained by having blocks or pieces of wood with holes suitably bored through them, and placed behind each other under the top.

Although only one screw and its fittings have been spoken of, it must be understood that the remarks apply equally to both. It may be well, however, to point out that on the end one the guide is fitted to the right of the screw. The thickness of the block also should be considerably greater than is necessary for the other, or its chief utility will be gone to such an extent that it might as well be dispensed with altogether. A thickness of 4in. will not be too great, though so much depends on the material that it is possible in some circumstances 3in. might be equally satisfactory; the reason for the extra thickness being that the stop (which will be seen by referring to Fig. 4) may be inserted in it without unduly weakening it. This stop, in conjunction with another in the bench-top, is one of the chief distinguishing features of the German bench, for by means of the two stops a board or other piece of wood can be securely held while being planed, &c.

It may be well at this place to say what is necessary about the stops and their arrangement. Along the top of the bench, at a distance of 3in. to 4in. from the front edge, and about 6in. apart from each other, a row of square (or, at any rate, rectangular) holes must be cut right through. In size they may be about ¾in. to 1in. each way, and they should be cleanly cut. A piece of hard wood, at least 1in. longer than the thickness of the bench-top, must next be cut to fit tightly into the holes. A similar hole and stop being in the end block, a piece of wood can easily be held between the two. The stop in the bench can be moved from one hole to another, as occasion may arise. When not required, the stops can be knocked level with the top and out of the way by a few taps from a hammer; they can be raised by the same means from below.

At the back of the bench the trough, or sunk tray, will be found very convenient for holding small tools, which would otherwise be in the way on the bench when working. It can be very simply made by just nailing a board of any suitable width (say 9in. or 10in.) near its edge to the under surface of the top. Another piece of board for the back and a piece for each end will complete the tray. As shavings and dust have a tendency to accumulate in it, it is not a bad plan to fasten a sloping piece of wood at one end to facilitate cleaning operations, for it is quite a mistake to suppose, as some amateurs apparently do, that a good workman is known by the quantity of litter and general mess which he makes. As a rule, the contrary is the case; the best workman is the neatest.

About the bench-top nothing more need be said, and the principal thing still to be done is to get it mounted and fastened to the legs. It might be rigid enough if fastened to the legs as they are; but it will be better to make sure of stability by means of the rails which are shown at the back and front of the bench near the ground, especially as the front one is exceedingly useful for another reason which will be mentioned presently. These rails may as well be of the same substance as the legs. The joint used should be the mortice and tenon or half-dovetail. If tenoned it will be rather an advantage than otherwise to let the tenon project an inch or so through the leg, and if the joint is not quite tight it can easily be made so by means of a wedge.

In Fig. 4 it will be noticed that there is a board perforated with holes between the top and the front rail. This board is intended to slide to and fro, so that when a peg of wood (not shown in the Cut) is inserted in the most convenient hole, it may serve to support one end of a board, the other of which is held by the bench-screw. This arrangement will be found specially useful when the edge of a long board has to be worked up for jointing, &c. To make a really good job the ends of this board, which need not be more than from 9in. to 12in. wide, should be clamped up, as shown in Fig. 10. This, however, is not absolutely necessary; but if it is not done, special care should be taken that the wood is thoroughly dry

FIG. 10.—ENDS OF SLIDING BOARDS.

and not likely to cast—*i.e.*, twist. The ends of the board may be let into grooves cut for them in the bench-top and in the rail at the bottom, but it will be better to cut them in the board. If this is, as it should be, not less than 1in. thick, though a little more will not be an objection, there will be ample substance to allow of a ½in. groove being ploughed. Two strips of wood, each nearly as long as the distance between the two front legs, must be prepared to fit easily within the grooves; one of them must then be screwed to the under-surface of the top, and the other to the upper surface of the rail. The board can then be moved as may be most convenient when working. Any number of holes can easily be bored through it with a centre-bit. It only remains to screw the top of the bench to the legs, or rather to the top framing connecting each end pair. If this framing is several inches deep, it may puzzle the novice to know how to get the screws through it into the top, unless indeed he uses screws of extraordinary length. The way is either to make a hole partly through the framing from below of such a size that the screw-head can be sunk a sufficient depth in it to allow of the point going far enough into the top to hold it fast, or to drive the screw through in a slanting direction from near the top of the framing. If this latter method be chosen it will, perhaps, be useful for the novice to examine the way a dining-table top is screwed to its frame. The chances are he will find that the screw-heads are sunk into sloping recesses, or pockets, cut for them, and he will have no difficulty in copying these on his bench.

FIG. 11.—DRAWER AND SUPPORTS.

In Fig. 11 a drawer is shown. It may be useful to have one to keep nails or small odds-and-ends in; but of course the bench is quite usable without it, and if the workshop is well found otherwise, the drawer will be of questionable advantage. In any case it will be apt to interfere with the free action of the sliding-board if at any time not pushed far enough in. The easiest way of fitting the drawer is suggested in Fig. 11, which is so clear that further explanation is unnecessary.

Some may think it will be an improvement to board over between the lower rails and if so there can be no objection to doing it ; but no special directions can be necessary.

Having now given all needed directions for making a thoroughly useful workbench it merely remains to be said if the maker cannot turn out well-finished work it will not be the fault of the bench, to which, if so inclined, he can add anything in the way of fancy stops, &c., that his taste may dictate.

TOOLS AND THEIR SELECTION.

THERE is as much special knowledge required in the choice of tools as in that of a piece of furniture, a scientific instrument, or a machine. And not only so, but the ability to keep them in working order involves knowledge which can only come after considerable experience. Hence, we propose to discuss not only the points of good tools, but the condition to be observed in order that they shall be maintained in the highest state of efficiency of which they are capable.

Taking the saws in the first place —sawing being preliminary to all other operations—there are these important points to be borne in mind : buckle, adaptability, forms of teeth, set, temper, handles. The various common saws are termed "hand," "tenon," "dovetail," "compass," "keyhole," and "bow." Our remarks will have reference to second-hand articles equally with those bought brand new from a shop, and some of the points enumerated seldom apply to new tools.

"Buckle" in a saw means that the plate is not in an equal condition of tension throughout, but that some portions are denser and more strained than others. The visible signs of buckle, when very bad, are bending of the blade, indentation, wind, irregular reflection of the light, and the inability to keep the saw straight in its cut. Holding a hand-saw by the handle, and vibrating the blade sharply, the presence of buckle will be detected by a flapping sonorous ring, sharp and pronounced, indicating that the saw plate is not in a condition of equal tension throughout. Buckle does not necessarily mean ruin to the saw, but it is difficult to remove, save by a skilful operator. Hence the best plan is to take the tool to a saw-sharpener, who will remove or minimise the evil by a process of hammering, effected on the self-same principle as the "raising" of works in sheet metal. As a matter of economy no one should buy a buckled saw second-hand, but it often happens that careless amateurs will buckle good saws by straining and forcing them to their work, and the attempt to hammer it out usually makes matters worse.

Of vital importance also is the adaptability, or otherwise, of a saw for its work. By this is not altogether meant the class of saw used—since the functions of the different kinds are obvious enough—but rather the shapes and setting of the teeth, matters so all-important that they will be treated in detail. The sizes and shapes of saw teeth are vastly modified according to the work they have to do, the difference consisting mainly in dimensions, "rake," "set," and "pitch." Size is partly dependent upon the class of saw, but varies much in the same type. Shape is governed by rake, being adapted to the work for which the saw is designed, while "set" governs the clearance of the saw blade in its cut, and on this depends the ease, therefore, with which it will pass through a piece of timber. The size of the tooth

Fig. 12.

A, Teeth of Ripping Saw ; B, Teeth of Dovetail Saw.

will vary from that shown in Fig. 12, A, which represents the teeth of a "rip saw," to those in Fig. 12, B, of a dovetail

Fig. 13.

A, Teeth of Cross-cut Saw ; B, Teeth of Ripping Saw.

saw. The rake varies from A, in Fig. 13, to B: these being the two extremes, A being adapted for "cross cutting," or

working across the grain, and, when of small size, for metal sawing; while B is only properly adapted for "ripping," or sawing down with the grain in soft woods. In Fig. 14, A represents the

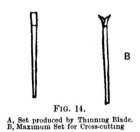

Fig. 14.

A, Set produced by Thinning Blade.
B, Maximum Set for Cross-cutting

minimum of set given, and B quite the maximum; the first being for metal sawing, and for hard and dry woods, and

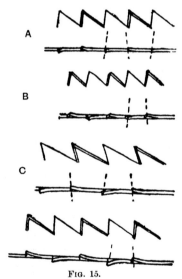

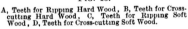

Fig. 15.

A, Teeth for Ripping Hard Wood ; B, Teeth for Cross-cutting Hard Wood ; C, Teeth for Ripping Soft Wood ; D, Teeth for Cross-cutting Soft Wood.

consisting only in a thinning down of the blade behind the teeth; the second representing the maximum for cross-cutting the soft woods, and consisting of a bending of alternate teeth to right and left. Now, in any case, the aim is to strike the happy mean both of rake and set for the particular class of work which any saw has to do, and to assist the amateur in this matter we here give the shapes of some typical teeth to actual sizes (Fig. 15, A, B, C, D), and from a study of these we may see that the general inferences to be drawn are: That the softer the wood the greater should be the amount of set, and of rake, and of "pitch" of teeth, or distance from tooth to tooth, while with the harder woods the opposite conditions obtain. The increased set is necessary in the softer woods, because their fibres do not lie so firm and dense as those of the hard woods; they tend, therefore, to press closely and cause increased friction against the saw blade; neither does their dust get away so freely from the cut, as the finer, drier dust from the hard woods. For this last reason also the pitch, or distance between teeth, is greater for soft than hard woods in order to give greater "spacing," since a saw with small tooth-space would become choked up with the fibrous, fluffy dust, which ought to get away immediately it is formed. For the same reason, also, a saw which will cut a piece of dry, well-seasoned wood with facility, will be found to operate with difficulty on the wood of the same kind when wet. Further, the rake, which in Fig. 15, C, acts so efficiently in penetrating and removing the material by its wedge-like action, is too penetrative for hard woods, hitching into the grain and jarring the hands of the workman. A saw having much rake is unsuitable for any cross cutting, for which work teeth having little or no rake and an excess of set are properly suited. Similar principles apply to the tenon and dovetail as to the larger hand-saws; but as a matter of practice, the typical tooth forms are often so modified that a kind of compromise is made, and a single handsaw or a single tenon saw is made to do duty on all kinds of wood and in all directions of the grain. Fig. 16 shows such teeth, which, though not ripping soft wood with such facility as Fig. 15, C, nor hard wood like Fig. 15, A, nor cross-cutting soft or hard woods like B, D, will yet operate on all but excessively wet stuff. In each of these figures the appropriate amount of set is shown in plan below the corresponding tooth profile. Wood which is very wet, as

some firewood logs and thick planking, can only be cut with a saw set specially

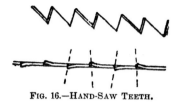

FIG. 16.—HAND-SAW TEETH.

wide and *regular;* moreover, a saw set and sharpened with uniform regularity is capable of doing a much wider range of work than one having more set imparted irregularly. This matter of setting and sharpening is of so much importance that we shall briefly describe the two processes.

Supposing that we have a saw which has been so badly used that the teeth are all "cows and calves," as the saying is (Fig. 17) which is very bad indeed.

FIG. 17.—SAW IN BAD ORDER.

The saw may be a very good one—good temper, elastic, free from buckle, and so on—but the teeth are spoiled, and we want to get these in working trim again.

In the first place, "top" the teeth, that is, run a file all down the points, the saw being held in a vice (Fig. 18) the while, until all the points are on one level,

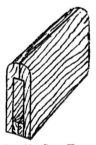

FIG. 18.—SAW VICE.

as indicated by the dotted line in Fig. 17. Then the teeth backs are to be filed until the points are renewed, and each tooth is not only on a level with its fellows, but sharp. This filing does not, however, take place straight across at right angles with the blade, but diagonally, as shown by dotted lines in plan (Figs. 15 and 16) the purpose of this being to reduce the friction of the tooth points, and to impart a correspondingly keener and more incisive cutting edge, it being an axiom that any cutting edge presented diagonally to material will cut with better facility than one presented at right angles thereto, because the former removes material in detail, beginning at one corner and working along to the other. The principle has its application in the diagonal bevel of a turning chisel, in the rounding of the edge of an axe or adze, in the "skew" rebate plane, in the operation of paring with a chisel moved diagonally, in the diagonal slicing of a knife, and in the slight angle given to the cutting edge of a saw tooth. In the saw the teeth are bevelled in alternate directions, as shown in Figs. 15 and 16, which balances the action, and prevents the saw from running to one side in its cut.

In practice all the teeth which slope in the same direction are sharpened in series, by which means the hand learns to preserve the precise angle alike for each, which would not be the case if its direction were reversed at each tooth. The saw, therefore, being clamped in the vice, every alternate tooth comprising the set which leans away from the operator is sharpened on the back. Then, the saw being reversed in the vice, the other half is also sharpened. With a saw in very bad order it is sometimes better to stop the teeth a second time after sharpening and touch them up once more with the file. The smaller the saw the more difficulty is there in sharpening and setting, and it becomes very troublesome in the smaller tenon and dovetail saws, the shape of whose teeth will rapidly become unequal by carelessness.

The "set" of saw teeth is variously given, and is of equal importance with correct sharpening. The best tools to use are the setting-block and hammer (Fig. 19). An automatic saw set is safer in the hands of an amateur, but workmen are not in love with such appliances, neither are they desirable from an educational point of view, since one of the chief uses of amateur handicraft is the training of the eye and hands. For those who care about automatic saw sets

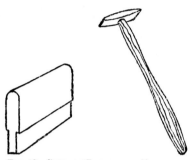

FIG. 19.—SETTING-BLOCK AND HAMMER

there are plenty in the market, most of them at reasonable prices.

To use the setting-hammer and block, proceed as follows: Screw the block in a vice and lay the saw teeth upon the rounding ridge, the blade running, of course, longitudinally (Fig. 20). The amount of tip given to the saw blade in relation to the block determines the

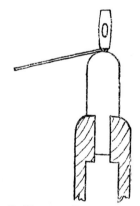

FIG. 20.—RELATIVE POSITIONS OF SAW, HAMMER, AND BLOCK.

quantity of set, and, once decided on, the slope must be maintained constant throughout the whole operation of setting. The saw being held in the left hand, the right grasps the hammer, and a series of rapid and accurately directed blows strike the teeth over until their further bending is arrested by the block. Turning the saw over and holding its blade at the same slope, the other half of the teeth are then struck. Always bear in mind that the set of the teeth is given in the same direction as the sharpening, to throw the cutting corners outside, as shown in Figs. 15 and 16. Another common mode of setting is that effected with the plyor set (Fig. 21). This is a plate of steel, notched like a wire or plate gauge, to take various

FIG. 21.—PLYOR SET.

thicknesses of saw plate. Each tooth in succession is pinched and bent over by a force of leverage gradually applied. This is not so efficient as the setting-hammer, because the elasticity of the tooth renders the precise and uniform setting of each less easy by this means than that accomplished by the sharp positive blow of a hammer. The plyor set is better adapted for small tenon and dovetail saws, and the hammer set for those of larger size.

If a saw is not set regularly it will not develop its full efficiency. If set regularly, but unsymmetrically—that is, if the teeth which lean in one direction have more set than those which lean in the opposite direction—the saw will "run" in the direction of greatest set. To ascertain if the set is regular, cast the eye down the saw teeth longitudinally, when any departure from truth will be at once visible. If the set is quite regular a needle should slide freely by its own weight down the middle plane of the teeth, the saw being held at a slight elevation from the horizontal. If the teeth bend and sharpen too easily, the saw is soft and will lose its edge rapidly. If they break when being set they are brittle, and each of these is an evil. There is no other mode of testing the temper of a saw blade than these—no mere inspection will reveal its quality; hence a shopkeeper should be required to warrant new saws, and exchange them if found faulty in these respects.

Handles thick and clumsy, or handles too thin should be avoided. Hands differ, so also should handles, and one ought to fit quite easily to the other, since weariness and-stiffness of muscle follow on the long continued use of unsuitable handles.

Concerning the names of the varieties and sizes of hand and tenon saws : a "rip" saw is simply a large hand saw, 28in. in length, and having about two-and-a-half teeth to the inch, and is scarcely required for amateur work. A "half rip " saw is the same length, but has four teeth to the inch. A " hand-saw " proper is 26in. long, and has about five teeth to the inch. This is the most generally useful for average work in carpentry and cabinet making. A "panel "saw is a small hand-saw, 20in. to 24in. in length, and is used for cutting the shoulders of tenons, the edges of thin wood, and so forth. Tenon and dovetail saws merge one into the other, and average from 8in. to 20in. in length. Those with iron backs are cheap, but those with brass backs are heavier and costlier. "Keyhole" and "compass," or "lock" saws are narrow, and apt to break off when forced to their work; they should, therefore, be used with care. The bow saw, being strained in its frame, is not liable to break; but when not in use the saw should always be relieved of tension by partially unwinding the string which imparts the strain thereto. The remarks made in reference to the sharpening and setting of saws apply equally to the small circulars used in the lathe, whose teeth are like those of the hand-saw in Fig. 16.

There are no tools more highly specialised than the planes, for these embrace not merely the simple jack. trying, and smoothing planes, for working level surfaces, but all the manifold bead, moulding, hollow, round, compass, and other types, for working irregular forms. We shall not attempt an enumeration of all these kinds—a toolmaker's catalogue (such as Melhuish's) will supply plenty of information of that sort—but simply indicate the points of good planes in general, and the way to preserve their full working efficiency.

For general use the following planes are to be considered as indispensable : Jack, trying, and smoothing. sometimes termed bench planes, because they are in perpetual request; rabbit or rebate plane, with skew mouth, small thumb plane; one or two plain beads; a selection of hollows and rounds, and a compass plane, all these being in wood. A plough is a useful addition, and absolutely necessary where grooving and tonguing have to be done, and may be picked up second-hand, being an expensive tool. For best work we would recommend at least one trying, smoothing, rebate, and bull-nose planes in iron.

A "trying" plane should be 22in. long, 2½in. iron; a "jack" plane, 17in. long, 2¼in. iron; a "smoothing" plane, 2¼in. double iron. Single-iron planes are not recommended for either of the above-mentioned. A "rebate" plane should be fitted with a guide and stop, allowing of any width rebate, say up to 1½in., the stop regulating the depth of same.

The stocks of planes are made of wood or of iron. Taking those of wood first, as being by far the commonest and most useful, there are two materials employed, beech for the larger, and box-wood occa-

FIG. 22.
WOOD UNSUITABLE FOR STOCK OF PLANE.

sionally for the smaller. In any case where wood is used, the grain should be straight, and cut at a considerable distance away from the heart. Thus Fig. 22 represents an unsuitable piece of

FIG. 23.
WOOD SUITABLE FOR STOCK OF PLANE.

wood, being near the heart, and Fig. 23 a good piece, far away from the centre

of the tree. The wood in Fig. 23 will remain true longer, and wear better than that in the first, and will not be nearly so liable to shake and warp with changes of temperature. Fig. 23, though with curly grain lengthways, would be better than Fig. 22 with straighter grain. The direction of the silver grain should be perpendicular or nearly so, as shown in Fig. 23; the face of the plane then remains smoother than when the rays run diagonally, as in Fig. 22. The material of the plane should be hard and heavy—a brown, mellow, close-grained wood wearing longer and working more sweetly than a light-coloured, open-grained timber. A saturation of the face with linseed oil tends to harden it and preserve its wearing capacity. The wood of a plane needs several years of seasoning before being worked up, otherwise it will twist, go out of truth, and crack with the warmth of a workshop.

Iron planes, of which many are now sold, should be of good material. They are made of cast iron only, and unless this is of the best quality, fracture will occur with rough usage. Iron planes are not adapted for what may be termed hedge carpentry or rustic work, such as many amateurs want to do. For rough, heavy, and outdoor work, the old-fashioned substantial wooden planes are more reliable, more durable, and will suffer less from accident. Iron planes are to be chosen only for the finer class of woodwork, such as cabinet-making and the best carpentry. Then they are. if kept in good order, splendid tools, working all ways of the grain with facility, and producing smooth surfaces; but they are all, when of good quality, costly, as a glance at the catalogues will show, so that a very moderate stock of such planes will cost a considerable sum of money, as compared with the expense of other tools and of wooden planes. Some few, as the trying, smoothing, bull-nose, and rebate planes, can be made by an amateur who is able to make the wood patterns and fit up the castings with blocking, but the task involves so much trouble as hardly to be worth the labour.

A new plane, however good, can, like a saw, be easily rendered inefficient through carelessness. Also a second-hand plane which will not work properly may, nevertheless, be a very good one.

Let us now consider, therefore, what are the conditions necessary to the efficiency of these tools.

Planes will work badly from three or four causes, and a workman who tries to maintain their efficiency at as high a standard as possible will sometimes have to give attention to each of these in turn before he can find the causes of their degeneracy, and rectify them. In the first place, since the value of a plane, as such, depends on its embodiment of what is termed the "guide principle," whatever produces inaccuracy in the face or sole must interfere with the correctness of its working. The soles of all planes, whether they have wood or iron stocks, lose their accuracy in the course of time because of wear, and in the case of the former, sometimes by the warping of improperly-seasoned stuff. In the case of an iron plane, want of truth can only follow on long-continued use, and then the file and scrape must be resorted to for the restoration of the original accuracy. With wood planes, the wear of a few months only is sufficient to necessitate the truing-up, or "shooting," as it is called, of the sole. This is performed as follows : First ascertain, by means of winding strips and straightedge, the precise extent to which the sole is winding, or rounding, or hollow, and also the localities of the highest portions. Keep the cutting-iron wedged in place, but knocked back about ⅛in. from its proper position, and screw the plane with its face upwards, in a bench-vice. Sharpen and set the iron of another trying plane very fine and very true transversely, and with it remove a few ribbon-like shavings from the highest portions of the sole of the plane which is screwed in the vice. Before reducing the whole surface, try the strips and straightedge, repeating the test three or four times if need be, the aim being to get as true a face as possible while reducing the thickness of the stock to as slight an extent as is necessary, because with each reduction in thickness the mouth of the plane is correspondingly widened. When the face shows a general dead level, even though a small patch or two of the original face may remain unremoved, it is wise to stop, because a localised depression here and there will not affect the general truth and accuracy of opera-

tion of the plane, while rapid widening of the mouth, due to excessive reduction of the face, is always an evil to be postponed as long as possible. After the face is "shot," saturate it with linseed oil, to smooth and harden the grain and impart a glossy skin.

After the maintenance of the truth of the bottom face, comes the grinding, sharpening, and setting of the iron. This, in most wooden planes, is double (Fig. 24)—that is, there is a lower or

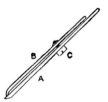

FIG. 24.—DOUBLE PLANE-IRON.

cutting iron (A) and an upper or non-cutting iron (B), the two being clamped together with the screw (C). The function of the top iron is the imparting of rigidity to the bottom one, by which its tendency to "chatter" is reduced, this tendency being inseparable from its mode of setting at an angle in the stock, while a true chisel should have its cutting face coincident, or nearly so, with the face of the material upon which it is operating. The top iron stiffens the lower, and prevents the vibration which would otherwise ensue, causing choking of the mouth, or a rough, uneven surface to be left by the plane. In some degree, also, it helps to prevent tearing up of the grain, by breaking and turning over the shavings immediately that they leave the cutting edge. Besides the setting of the top iron, results are affected by the mode of grinding and sharpening of the cutting iron, the bedding of the latter on its seat and wedging of the same, and the width of "mouth," or opening in the sole, through which the iron projects.

The cutting-iron may be ground and sharpened either too thin or too thick. In the latter case, its proper wedge-like action will not come into play, and the surface of the material will be rubbed, scraped, and cut in turn, much labour being expended with little good result.

If, on the other hand, the iron is ground and sharpened too thin, it will not retain its edge, but after the removal of a very few shavings, especially in harsh, hard, and knotty stuff, it will become finely notched and leave grooves and scratches over the surface. In each extreme also the tendency is to cause the shavings to choke in the mouth. Hence the happy mean is to grind and sharpen at such angles that the iron will not only cut sweetly and freely, but retain its edge for a reasonable length of time. These angles should depend to some extent upon the nature of the material being operated upon, whether hard and cross-grained, or soft, silky, and straight-grained, the angles being increased for the first and diminished for the last. Fig. 25 shows the ground and sharpened

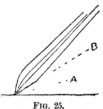

FIG. 25.
ANGLES FOR GRINDING AND SHARPENING.

angles of a trying-plane iron suitable for soft wood and hard wood alike, though not adapted for very hard and crooked curly grain. Of course the sharpening angle (A) will be constantly changing, being first but slightly greater than the grinding angle (B), but gradually thickening with each re-sharpening, until the facet becomes almost or quite coincident with the face of the material, at which stage re-grinding becomes necessary.

The position of the top iron should properly vary with different classes of material. The harder the wood, the greater tendency there is to chatter and choke, with the consequent production of a roughened and wavy surface; and the top iron (B in Fig. 25) should therefore be brought very close to the cutting edge, so that in certain cases the edge of the cutting-iron (A) will project only as a fine line beyond the top iron. In soft woods, on the contrary, the best results are obtained by keeping the top iron farther back, say to the extent of $\frac{1}{32}$in. full, or even $\frac{1}{16}$in. if the wood is wet, a distance quite unsuited to hard curly stuff. Fig. 25 shows the top iron placed at a medium distance back, adapted for ordinary working of dry pine and deal, or straight-grained hard wood.

The proper bedding of the iron on its seat, and its even wedging, have much influence in preventing choking of the shaving. If the iron rocks upon its seat, though only in a very slight degree, or if the wedge does not exert an equal pressure all over alike, choking will certainly occur, and these points must therefore be looked to if the plane is found to work unsatisfactorily; they are easily remedied by removing the higher points of contact of the wood, with a chisel in the case of the seating, and with a plane when the wedge is tight. The wedge should never be tight sideways, since, if then driven in hard, it is apt to split out the wood of the stock close beside the iron. A very slight correction of the bedding of the iron and the fitting of the wedge will often work marvels.

The width of the mouth ought not to be greater than will permit of the free passage of the shavings upwards, notwithstanding that, in consequence of the wearing of the face, it becomes unavoidably increased. A moderate increase in width, say up to $\frac{1}{8}$in., will not materially interfere with the proper working of the plane; but anything exceeding that amount interferes with the proper planing of short pieces of stuff, causing a slight digging inwards of the iron to occur just at the commencement of removing a shaving. Then the usual practice is to reduce the mouth to its original width, by fitting in a dovetailed slip of wood in front of the cutting-edge of the iron. Some even glue a fresh sole, $\frac{3}{8}$in. or $\frac{1}{2}$in. thick, on the face of the plane, throughout its whole length, thus narrowing the mouth, and at the same time restoring the plane to its original depth and weight, both being points of value, a heavy bench plane always working better than a light one.

The widening of the mouth is partly affected by the fact that all common plane-irons taper backwards, being thinner at the top end than next the cutting-edge; the iron, therefore, as it becomes worn back, leaves the mouth more open in consequence. But there are parallel or gauged irons made, at the cost of a few pence more, by the use of which the evil of widening of the mouth is lessened.

The grinding of plane-irons should be properly done without the aid of artificial means; but whoever can grind a broad trying plane-iron straight across, producing a single facet only, need not be afraid to tackle any other tools. The difference between grinding wide and narrow tools is that the former are more difficult to keep in one horizontal plane against the stone, the tendency of the latter as it revolves being to thrust the iron downwards, with the consequent production of a fresh facet or facets. Practice will soon overcome the initial difficulty. The iron should be ground until the sharpened edge is nearly but not absolutely effaced. Irons vary in curvature in the transverse direction. That of the jack plane should have from $\frac{1}{32}$in. to $\frac{1}{16}$in. of rounding, suitable to the rough class of work which it has to do. But the trying and smoothing plane-irons should be almost, though not quite, straight across, since their function is to true and finish broad flat surfaces, whereas if they were rounding they would leave hollows in the face. The best test for the truth of the irons of the trying and smoothing planes is to place the edge along the face of the plane itself, the iron standing perpendicularly thereto, and to note the coincidence between the edge and the face, which should be almost absolute, the light barely showing between the two near the extreme edges. Then the keen corners are just rubbed off on the hone to prevent the marking of grooves thereby. All the actual rubbing on the hone for the purpose of sharpening must be done on the bevelled facet, never on the face, the only object of turning the face over upon the hone being to thrust back the "wire edge," and so loosen and remove it.

The irons are removed from jack, trying, and panel planes (Fig. 26) by striking a sharp blow upon the top of the stock near the end in front of the wedge. The smoothing-plane (Fig. 27) is struck on the hinder end in each case, the removal being effected by the reaction of the blow. Battered planes look very unsightly, hence the hammer should be applied so that its broad plane and not the edges shall strike the wood, and the wedges should not be very tight, since one, or at the most two, sharp blows should be sufficient to

FIG. 26.—JACK OR PANEL PLANE.

release the iron. Bead, moulding, hollow and round planes are released by striking beneath the shoulder at the upper end of the wedge.

The amount of projection of irons beyond the face is gauged by taking a sight down the face, when it is easy to see, not only whether the projection is correct in amount, but also if the irons

FIG. 27.—SMOOTHING PLANE.

stand square and equal across. Minute adjustment is made by tapping the stock with the hammer to diminish projection, by striking the iron to increase it, and by tapping it sideways to bring it parallel and square across.

These, then, are the most important points which concern planes. Considering how valuable these tools are, being in perpetual request, it is not to be wondered at that workmen and competent amateurs should take pride in maintaining their planes in proper efficiency.

The quality which is of most importance in edge tools is temper. This cannot be known except by trial—that is, by grinding and sharpening, and applying the edge against hard, harsh, and knotty stuff, to learn how it stands. A tool which is too brittle will become

C

notched : the edge of a soft tool will turn over almost immediately. A hard, even, though somewhat brittle, tool is decidedly preferable to a soft one, because by grinding its angle rather large or "thick" it will cut very well, while a soft tool will not keep its edge under any conditions of grinding. The quality of a cutting tool cannot always be estimated from its first grinding, since it may be unsatisfactory then, and improve after $\frac{1}{8}$in. or $\frac{1}{4}$in. has been ground off; hence a fair trial should be given by grinding back to that amount.

Next to temper, there should be accuracy of profile. A paring chisel or gouge intended to be straight should not be crooked, since the face ought to be, in large measure, a guide to true cutting. A gouge, moreover, should be of a regular sweep—not of varying radii, flat and quick alternately—and a chisel-face should be so flat across that, when it is rubbed on the hone, contact should occur all over the surface, showing that it is not even slightly hollow or rounding.

Finish—that is, polishing—counts for little ; but as a rule the best tools have the best finish. Though a well-finished tool may be badly tempered, one that will take a high polish can scarcely be made of inferior steel.

Flaws are not of rare occurrence, being caused by the hardening process. They can seldom be detected until the tool actually breaks ; but if fracture is due to the presence of a flaw, its locality and extent will be apparent by a dark spot, like that produced by the stain of nitric acid. Then, if the tool is taken to the salesman, he will, if it has been warranted, exchange it for another.

The mode of grinding chisels and gouges is varied slightly, according to the nature of the work which they have to do. Thus, a tool ground for use with the mallet should not be so keen as one employed for paring only, and a tool for hard wood should not be quite so keen as one for soft wood, durability of edge as well as cutting capacity having to be borne in mind. No tool will cut well unless it is kept perfectly flat upon the flat face, hence this is rubbed with the gouge-slip or upon the hone, as the case may be, only just sufficiently to turn back the burr formed by sharpening on the bevel.

The brace and bits are tools which are quite indispensable. A brace and a dozen bits can be bought for 5s. or 6s., and better sets at prices ranging up as high as 30s. or 35s. The quality of the bits is pretty much the same, the difference in cost being mainly in the brace and in the number of bits to the set. And there is this difference in bits— they may be black, bright, or straw-coloured, the first-named being the cheapest and the last-named the most costly. We do not know that there is any difference in temper, though the straw-coloured kind are usually believed by workmen to be superior to the others. A complete set numbers from thirty to thirty-six, and embraces centre, nose, shell, countersink, and some others of a miscellaneous character. There is seldom anything wrong with these when bought new ; but in the cheap sets the bits are often not fitted to the spring catch of the brace, which is a point that should be noticed, since when not so fitted the bit cannot be withdrawn along with and after the brace from its hole, but has instead to be pulled out by hand or with pincers. With an unfitted set, either file the notches yourself in all the bits alike or buy a common iron brace with a pinching screw only. The temper of the bits is necessarily very soft in all cases ; highly-tempered tools would not stand the torsional stresses to which they are subjected. For sharpening these, a smooth file alone, or a file followed by a gouge-slip, is used.

Brace-bits will soon get altogether out of order by careless sharpening. Just as we should not sharpen a chisel upon

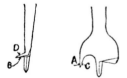

FIG. 28.—CENTRE-BIT.

its flat face, neither should we sharpen a centre-bit on the outer edge of the nicker (A, Fig. 28) or on the lower face of the cutter (B), but only on C and D. If we sharpen on B we soon alter the cutting angle : if we sharpen at A we reduce the radius of the bit, with the result ultimately of making the radius of the nicker equal to, or even less than, that of the cutter, whereas it should always be $\frac{1}{32}$in. or $\frac{1}{16}$in. greater, in order to bore a clean hole. For a similar reason we should never get the nicker as shallow in the horizontal direction as the cutter, or shallower, but always maintain it projecting $\frac{1}{16}$in. or $\frac{1}{8}$in. beyond, in order that it may first enter and divide the grain-fibres, so that the cutter may remove the inscribed portions in the form of minute shavings or chips. The nicker and cutter fulfil distinct functions, and the relations which they bear to each other must be preserved intact, just as when purchased, in order to ensure the proper performance of those functions.

The centre-bits, however well sharpened, bore with difficulty in end or diagonal grain, hence it is good policy to incur the extra outlay involved in the purchase of the "auger bits" with two nickers and two cutters, preferably of the Jennings type. These are made from $\frac{1}{4}$in. to 1in. in diameter, advancing by sixteenths. They average from 1s. 6d. to 2s. each, but will bore in any direction of the grain perfectly clean and true, and quite straight.

For those whose work is not of a rough character, and who require to bore holes to divisions of the inch more minute than $\frac{1}{16}$in., Clarke's Expansive bit (Fig. 29) is extremely handy. These

FIG. 29.—CLARKE'S EXPANSIVE BIT.

are capable of considerable range in diameter; two bits, each with two cutters, embracing all sizes from $\frac{1}{2}$in. to 3in. The graduations read to $\frac{1}{32}$in.

Nose bits should not be sharpened upon the outside edge or upon the face of the cutter. The proper places to sharpen, using a fine file, are on the inside faces (A A, Fig. 30). It is seldom these faces need sharpening at all, and the less often it is repeated the better will the tools preserve their forms and functions intact. The same remark applies to gimlets and augers.

The only necessary test of a square is its being truly at right angles, which is tried by laying the stock against the perfectly straight edge of a board, and

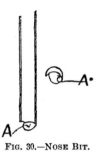

FIG. 30.—NOSE BIT.

describing a line along the edge of the blade upon the board. Then on turning the square over, the want of coincidence of the blade with the line scribed will indicate twice the amount of divergence from the true right angle. Squares are very often out of truth a trifle, and should by no means be condemned for that reason, since a little careful filing will set them right. They get out of truth by constant wear also, and should on this account be periodically tested. The general finish of a new square is usually a fair test of good quality. Thus, the blade should be at right angles to its stock, instead of running slightly diagonally; the brass facing of the stock should be neatly put on, and closely fitting; a roughly-made square, like many other articles, will usually be found wanting in strict accuracy also.

The temper of screwdriver-points should be such that they neither bend nor break under stress, which quality can be tested on a screw well rusted in. Some screwdrivers are badly handled, clumsy handles having sharp angular edges which make the hand sore in using them. The handle should be flattened of course for leverage, but the edges should be neatly rounded: they then operate well without tiring the hand. Fig. 31 shows a handle of most approved form. Screwdrivers made of round steel like that in the figure are the handiest for general work, because of the facility with which they enter deep centre-bit and countersunk holes.

c 2

Pincers should be broad, and rather flat across the faces when the jaws are closed. If the faces are much rounded this means a distinct sacrifice of leverage.

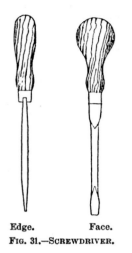

Edge.   Face.
FIG. 31.—SCREWDRIVER.

For the same reason small pincers are not so useful as the larger ones. Pincers are only steel-faced to perhaps $\frac{1}{16}$in. in thickness, and are not always what they should be, often being either so soft that the jaws become indented, or so brittle that the edges chip away. When first bought they should be tested on some stiff nails, and if unable to stand the stress should be returned.

The only test of a hammer is the durability of the face, which may be soft on the one hand, or brittle on the other. The size of head depends on the class of work being done. Two or three hammers of different weights are useful. The handles should be long rather than short.

Mallets should be made preferably of beech or ash, and of the hardest, heaviest, and straightest-grained wood available. The taper of the handle in the eye alone retains the head from flying off. A slight chamfer of the edges is an excellent preservative against too rapid splitting of the head.

General finish is the best test of the excellence of quality of compasses and dividers. The black ones may be quite as good as, or better than, bright ones—in fact are often costlier. Large numbers of cheap compasses and dividers are utter trash, lacking in all those qualities which these tools should possess. Joints should be well fitted, screws deeply and cleanly cut, fitting closely in their tapped holes, proportions should be good, and stiffness is essential. The arch or spring of dividers should be stout and strong. The best compasses are those in which minute and precise adjustment of the wing is provided for by means of a wing-nut, which operates after the pinching-screw is set.

Metal trammels of good quality are rather costly, and the cheap ones lack stability on their rods; but a cheap trammel can be made by turning a couple of heads in boxwood (Fig. 32), cutting through the slots with a mortise-chisel, and tightening them upon the rod with

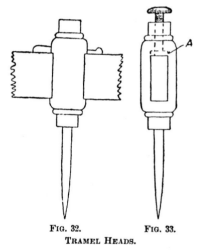

FIG. 32.   FIG. 33.
TRAMMEL HEADS.

wedges, as shown. For those who have suitable taps and dies, the better plan is that of Fig. 33, where a small nut is let into the head of the trammel, at A, the nut being tapped to take the pinching-screw. The pinching-screw is tightened on a brass washer-plate interposed between the head and the trammel-rod. The points in each case are of steel, driven tightly into the wood, which is banded with ferrules in order to prevent splitting out of the grain.

Rules and scales may be had of almost any quality and price, from 6d. to 30s.

each, and they should therefore be chosen altogether with reference to the class of work which is likely to be done. Considering that the wood has to season for about seven years, and that expensive machines are required for marking off the divisions, cheapness can hardly be looked for in articles of this character. A good boxwood slide-rule suitable for shop use can hardly cost less than from 4s. to 5s., but at this price it is quite reliable. The best scales are those which are open-divided—that is, instead of the smaller sub-divisions being carried right along, the two end ones alone are minutely sub-divided, which prevents confusion in the reading off of dimensions. The universal scales, in which the whole surface is covered with divisions, are a plague, as being wasteful of time; hence, those scales alone are of much service which are divided along the edges only. A coat or two of shellac varnish or of French polish is an excellent preservative for rules and scales, keeping them clean, and thus preserving their divisions distinctly visible, even though in perpetual use.

Grindstones and hones are numerous, and in each the test of quality lies in trial. The first-named may be either natural or artificial, it matters not which. The larger the stone, the easier the work of grinding, and the greater the economy of time. Perhaps the cheapest way is to buy the stone and rig it up with treadle and trough one's self. The main difficulty lies in hanging the stone on

FIG. 34.—GRINDSTONE (MOUNTED).

its spindle so that it runs true. It is accomplished by adjusting wedges, the hole in the stone being perhaps ½in.

larger than the bar on which it is fitted, so affording play for such adjustment. A grindstone should be mounted to work with treadle, so that both hands are left free to hold the tool when grinding. Fig. 34 will give the amateur an idea of a mounted grindstone with foot-power. The stand is very easy to make; to buy one ready-made, with a 16in. or 18in. stone, would cost from 25s. to 30s., whereas a home-made one need not cost more than 10s., using either a 16in. or an 18in. stone, the only difficult part being blocking the spindle into the stone, so that it will run true. For an 18in. stone, the size of trough should be 22in. long, 11in. high (inside measure), the width entirely depending on the width of the stone, but leave at least 1in. clear each side between it and the sides of trough.

The trough should be made of deal, not less than ¾in. thick. The sides should be grooved ¼in. deep for the ends to fit into (Fig. 35), as the trough must necessarily be watertight, and this will

FIG. 35.—SIDE OF TROUGH, SHOWING GROOVES FOR THE ENDS.

help to make it so. All parts to be screwed together, not nailed. It will be seen by referring to Fig. 34, that one end is much longer than the other; this is to prevent the water from spilling, as in grinding the wheel revolves from the operator. An average height for the stand is 2ft. 6in., the legs to be of deal quartering 2in. square, a shoulder being cut 3in. from the top, ¾in. deep, on which the bottom of the trough rests. The bottom of the legs should project rather beyond the top of the trough to ensure firmness. The rails between the legs are 2ft. by 1½in., and are mortised into them. The stretcher from rail to rail is the same size, the length of rails and stretcher depending upon the size of grindstone to be used. The iron spindle (Fig. 36) can easily be made by a blacksmith. By referring to the drawing it will be seen that the centre part of the spindle is left square, say ¾in., and a knob is forged at the end of the crank to prevent the treadle-rod from

FIG. 36.—IRON SPINDLE FOR GRINDSTONE.

slipping off. The treadle and connecting-rod are both of wood; the tread 3in. by 1in., with length accordingly; the rod 2in. by ¾in.

To mount the stone, prepare several small wooden wedges (four for each side, as the hole in the stone is square), and knock them in evenly until the spindle is properly centred; the stone should run on iron bearings fixed on the sides of the trough, which can also be made by a blacksmith. A simple plan is an iron plate with a staple riveted into it, the plate being fixed by screws on to the edges of the sides. With the aid of a little oil the spindle will revolve easily on the plate, the staple keeping it in position. The bottom of the trough should be screwed on, and the whole thing painted inside and out; in case of leakage put a little putty round the bottom inside. The amateur will find what great advantage a foot-power grindstone has over a hand one, for in grinding a plane-iron or chisel the tool must be held firmly in order to procure a proper edge, and a good idea is to fix a small wooden rest on the front end of the trough to lay the tool on when grinding.

Stones at rest should never be allowed to stand in water in the trough, since that tends to soften them and causes them to wear more rapidly in the softer portions. When they wear into grooves, it is necessary to true them up by turning, a bar of steel about ½in. or ⅝in. square, drawn down to a point at one end, being used for the purpose, and continually turned around to present new sharp corners to the stone. The face of the stone should be turned about $\frac{1}{16}$in. or ⅛in. rounding.

Of the hones, or oil-stones, the best are undoubtedly Turkey or Charnley Forest, as being more reliable than Washita, Grecian, or Nova Scotia. The quality of Turkey is not so uniform either as that of Charnley Forest; for though among the latter some are better than others, we never yet met with a bad or a poor one. They are cheap also, costing only about 8d. per pound, and a stone of 3lb. weight is quite large enough. A hone is mounted in a wooden stock, with a cover to protect it from dust, and will last the best part of a lifetime. Get a piece of deal quartering, 3in. square, and about 3in. longer than your stone. Saw the wood in halves, lengthwise, and let the stone about ½in. into one piece,

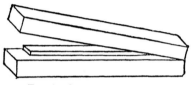

FIG. 37.—OILSTONE (MOUNTED).

the other forming the lid or cap to keep the stone free from dust and grit (Fig. 37). When it wears hollow it must be rubbed down level either upon a piece of sheet iron with sharp sand, or upon a sheet of coarse emery cloth.

We may perhaps include in this list the bench-stop. That shown in Fig. 38 is an old-fashioned type, but much used even now by regular carpenters. It

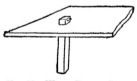

FIG. 38.—WOOD BENCH-STOP.

consists of a piece of wood, 2in. square and about 12in. long, fitted vertically into the bench, say 6in. from the top end and front. This is fitted tightly, so that it can only be raised or lowered by hammering, and in one side, close to the top, two or three short iron spikes are driven in, projecting ¼in., to hold the wood that is being planed. The stop should be made of beech or oak, or any hard wood. The iron spikes may be made by driving in French wire nails, leaving about ¼in. showing, then filing off the heads, and pointing them like a bradawl; but, as I have said, they should only project when finished about ¼in. Fig. 39 shows the ordinary

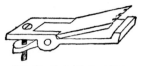

FIG. 39.—IRON BENCH-STOP.

pattern iron bench-stop, which is let in flush with the top of the bench, and the toothed stop-plate can be raised by turning the screw at the back, as shown. These can be purchased at 1s. each. A still more improved pattern is shown in Fig. 40. The plate is fixed flush in the bench, the stop being raised or lowered by giving half a turn to the screw in front. When set for use it is as firm in any position as the bench

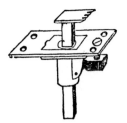

FIG. 40.—THE "PERFECT" BENCH-STOP.

itself, having no vibration at all. It is rightly called the "Perfect" bench-stop, and is sold by Messrs. Moseley and Son, 323, High Holborn, London, W.C., the price being 2s. 6d.

Another useful addition to the workshop that the amateur should make is a

FIG. 41.—SHOOTING-BOARD.

shooting-board, as shown in Fig. 41. No definite size can be given for this, though it is as well to make two, one for shooting the edges of narrow wood, such as in making picture-frames, the other for wider wood.

For the sake of describing, we will presume the bottom board to be 12in. wide, 1in. thick, the edges planed perfectly true, the top board 6in. wide, 1in. thick, any length. In the bottom board are fixed two bolts with square shoulders, so as to prevent their turning round, 2½in. by ⅜in., with flat heads, let in flush with the wood. These should be fixed about 3in. from the side, and at a proportionate distance from the ends, according to the length of the board. In the top board two slots are cut sufficiently wide to allow the bolts to slide freely, the length of the slot to be within 1in. of each side of the board—*i.e.*, in a 6in. board there will be 4in. slots, so that the width of the shooting-board can be regulated to anything within the length of the slots. The top board is then screwed to the under one with nuts, an iron washer being fixed between the nuts and the wood. At the top end of the bottom board, a wooden stop should be fixed for the wood to be planed to press against.

The trying-plane is used for shooting edges, as is well known, and in this case it is used by being put on its side. When it is required to join two pieces of wood, by shooting the edges with a board as described, the amateur cannot fail to make the join true all along.

Other useful additions can be made by the amateur, such as a bench-dog and wood cramps. Fig. 42 shows a bench-dog. It acts as a stop for holding

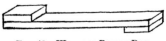

FIG. 42.—WOODEN BENCH-DOG.

wood against when sawing with the tenon-saw. The appliance should be about 10in. long, 3in. wide, and ⅝in. thick. A wooden stop, 1½in. wide, is screwed on either end, the one the

FIG. 43.—WOOD CRAMP AND WEDGES.

reverse side to the other. Wooden cramps for holding glued joints are easily made, and the amateur should possess at least two. It is as well to make them of moderate length, and by having one stop, movable boards of various lengths can be held in them. A cramp like Fig. 43 might be 2ft. 6in. long, 4in. wide, and 1in. thick. The stops should be 1½in. thick, 3in. wide; one should be fixed and the other movable. Two wedges are used for each cramp.

The things which the worker may as well make for himself are a mitre-block or box and a mitre-shoot. Though there is not much to choose between the merits of the block and box in the hands of an expert, the novice will probably find the latter the more con-

FIG. 44.—MITRE-BOX.

venient form of the two. In it the saw is guided both in front and behind the piece of moulding which is being cut, while in the other it is only guided in one place. The box alone need therefore be described. An ordinary form is shown in Fig. 44. It consists merely of three pieces of wood one of which forms the bottom, and the others make the sides, the ends and top being open.

About 18in. may be taken as a suitable length. The height of the sides on the inside must not be greater than the width of the saw from its edge to the back rim, and if the width of the bottom is sufficient to allow of the moulding being placed within the sides, it will be enough. Pine about 1in. thick will form a suitable material.

The bottom inside must be level, and the angles square. The saw-cuts which will form a guide for the saw through the moulding must be absolutely perpendicular from top to bottom, each pair across being in the same straight line, so that the saw works through both sides without binding.

The cuts must be accurately made at an angle of 45deg. transversely. There are many ways in which this angle can be accurately ascertained and marked on the wood; but perhaps the simplest which can be suggested for anyone who has not a mitre-template, or square with one side set at 45deg., is as follows: Take a piece of card and make a perfect square, each side of which is not less than the width of the box across. Divide the square piece into two equal triangles by cutting it across diagonally. Either or both of these may then be used as a guide to the direction of the saw-cuts by laying them on the top edges of the box.

The shoot (Fig. 45) consists of a piece of board on which is fastened another with a straight edge. On this latter is another with its sides placed accurately at the same angle as the cuts in the box. As before, the wood may be pine, and the size anything that is convenient. The top mitred-block serves as a top against which the moulding is held while being shot, and the straight edge

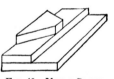

FIG. 45.—MITRE-SHOOT.

of the narrower piece forms as a guide for the sole or bottom of the plane, the side of which works on the bottom board.

There are other tools which have not been enumerated, but these are the ones in most common use, and the hints here given will also apply in the main to those not formally included in the list. A workman can judge of a good tool almost as if by instinct; but the case is different with amateurs, and it is to guard such against the purchase of some of the cheap trash which is sometimes sold that this advice has been given. It should not be forgotten that good tools, like most other commodities, have their price.

HINTS ABOUT WOOD.

THE amateur worker in wood, whether he be carver, tur ner, fancy cabinet-maker, or plain joiner, seems to be in a constant state of trouble with the material, if one may judge from his frequently expressed desire to know where he can buy seasoned timber. Now, with a view to help him, we give a few common-sense hints. If he acts upon these, the probability is that the difficulties arising from the use of improperly-prepared stuff will be matters of the past. At present he is rather too apt to think that as he is "only" an amateur, he has badly-seasoned wood palmed off on him. That this may occasionally be the case I am not prepared to deny; but I think it may safely be assumed that anyone buying from a respectable man will not have badly-seasoned wood sold to him instead of something better. What then, it may be asked, is the reason why boards bought from the timber merchant, and guaranteed to have been so-or-so many years seasoning, crack and twist after they have been worked up and subjected to the ordinary atmosphere of a dwelling-house? If the wood was really seasoned, why should it crack when made up and put in a warm room? Of course, the average amateur jumps at once to the conclusion that the timber merchant has taken advantage of his inexperience. But that is not necessarily so.

Perhaps the principal error that is made is in supposing that seasoned timber is fit for immediate working up. Occasionally it may be, but very often—perhaps generally—it is not. Without dilating on the actual seasoning of timber a few remarks on rendering fairly well-seasoned stuff fit for use will be of value. First of all, it should be remembered that timber shrinks, owing to the evaporation of the moisture it contains. In other words, as it becomes dry it shrinks. If a board be allowed to dry before it is fastened at the edges it is able to contract without splitting. If, however, it is bound at the edges to something which will not yield, it is almost certain to split. We now have the clue

to the proper treatment of wood befor e working it up. As it comes from the timber-yard it is, even if sound-seasoned stuff, more or less damp. The natural sap of the wood may have long ago been got rid of, but a certain amount of mois ture is in the wood. This moisture is practically unavoidable, as wood is absorbent, some kinds more than others, and unless kept in a perfectly dry place it is sure to be slightly swollen with damp received from the atmosphere.

It is, of course, absurd to suppose that timber can always be kept in a warm, dry place, and knowing this, a careful, practical cabinet-maker would hardly risk his reputation by using stuff fresh from the yard. Instead of doing so he will keep the wood in a warm, dry place for a space of days, hours, or weeks, according to circumstances, in order that it may be thoroughly dried and shrunk before it is worked up. The amateur should do the same. If he does, so-called unseasoned timber will be a thing of the past.

Amateurs who want their wood ready for immediate use can always obtain it from a good manufacturing cabinet-maker. The price will, however, be much higher than if got at a timber-yard. It may, indeed, almost be said that dry wood fit for immediate use can be got at any time if purchasers are willing to pay the price; but that the timber-yard is not the place to buy it at. Stuff bought from the yard should, if only as a precautionary matter, be kept in a warm place before it is worked up. What the place is must depend on the state of the weather. In this climate an outhouse is not suitable, except perhaps during a prolonged spell of dry, hot weather. Neither will an ordinary cellar do. Timber might season in these places, but it certainly will not dry. It must be kept in the driest and warmest place in the house, and naturally the kitchen will occur to most amateurs as the very place they want as a drying-room. The wood should be gradually, not suddenly, dried; therefore it should not be subjected to immoderate or sudden heat.

The best plan will be to put the wood in some out-of-the-way corner, and if there are several pieces they should not be placed in close contact with each other. The warm air should be allowed to get to all equally; so that the moisture may be taken from all of them and not merely from the outside board. To ensure equal drying the relative positions of the boards should be changed now and again. They may be laid, but it will be better to let merely the ends rest on the ground. Of course this applies to boards, as this is the most usual form in which wood is bought; but in the case of small bits, such as odd pieces for carving or turning, the worker must use his discretion as to the best position in which to dry them.

Some kinds of timber are extremely difficult to dry properly, for though the pieces may seem all right till they are cut or worked up, as soon as a fresh surface is exposed to the air they exhibit all the vagaries of new, unseasoned stuff. This simply means that extra precautions should be taken with timbers that are known to be difficult to bring into a workable condition. Fortunately, none of those most commonly used by amateurs require this extra care—pine, American walnut, ash, mahogany of all kinds, and ordinary oak, being easily treated. It is unnecessary to name those which are not reliable, except to say that some kinds of oak, especially the English brown or pollard variety, and pitch-pine require care. As a general rule, it may be stated that timbers which have sudden alternations between hard and soft figuring require more care than others of a more uniform texture. This, however, it must be quite understood, is a rule of general application only.

It follows from what has already been said, that all furniture should be made up in dry, warm rooms, and that unheated outhouses or cellars, in which so many amateurs are compelled to follow their hobbies, are not suitable places to work in. The timber swells up, possibly not to any great extent, but quite sufficient to make it "go " (wrong) when removed to a warm room. Especially in winter time should care be taken that wood is thoroughly dry.

If anyone wants to experiment with the shrinkage powers of wood from which all moisture has not been removed,

let him fit a piece accurately to any opening in an article of furniture. A drawer-space will do very well. The drawer may be temporarily taken out and the experimental piece fitted to the place the front of the drawer usually occupies. Remembering that wood does not shrink to any appreciable extent in length—i.e., in the direction of the grain—it is only necessary that the fit should be exact at the top and bottom. After this has been done, put the piece in a warm, dry place for a time, and when it has become dried put it in the opening to which it had been previously fitted. The chances are that it will have shrunk to such an extent that, had it really been intended for the drawer front, a badly-fitting drawer would have been the result. For this reason new drawers are usually made to fit very tightly, as experience shows that there will very likely be subsequent shrinkage. If they do not shrink they can be "eased."

While on the subject of drawers it may perhaps be well for the benefit of those who did not understand what was meant by binding wood in such a way that it could not contract without splitting, to explain here further, as a drawer bottom affords a familiar example. Few who have had anything to do with new furniture will have failed to notice that the bottom of a drawer frequently recedes from the front, leaving a slight opening ; this is owing to the shrinkage of the bottom, which is usually fastened to the back. Now if the front edge of the bottom were fastened by glue or other means into the groove within which it loosely fits as well as at the back, the natural shrinkage would cause a split. It will thus be seen that not only drying should be attended to, but that the natural tendencies of the material used should be known and understood. In this (apart from finish) consists the proper construction of cabinet furniture, and it is not too much to say that it is a matter very imperfectly understood, not only by amateurs, but by many of those engaged in the furniture trade.

Reference to shrinkage of timber has been referred to, but it must not be forgotten that as dryness causes wood to shrink, so does damp cause a perfectly dry piece of wood to swell. This, however, is of such comparatively small importance to the worker that space need not be devoted to its consideration. It may be well to mention that wood is always liable to expand or contract, as the case may be, with variations of temperature, and that, take what care one will, it is absolutely impossible altogether to avoid mishaps. We could, did space permit, mention many instances which have come within our own experience, but must refrain, for obvious reasons.

In conclusion it may be stated that when wood has been French polished or varnished it is less likely to change than when " in the white," for the simple reason that the grain is closed.

HINTS ABOUT GLUE.

TO the worker in wood, whether amateur or professional, as well as for ordinary domestic use, glue is such an indispensable material that no apology need be offered for directing attention to it, though to many the subject may seem a somewhat profitless one. Unfortunately, though glue is in such constant use, its right preparation and application often receive insufficient care, especially at the hands of amateurs, whose work is confined principally to mending any little article that may have got broken. The thoughtlessness with which one so often sees glue used can only be because its proper manipulation is not properly understood.

First of all it goes without saying that there is more than one quality. Good glue can be had from any respectable dealer, but let us here utter a word of caution against low-priced glues: they are seldom to be depended upon, and the difference between top and bottom figures, is when one takes into account the long way a pound of glue will go in ordinary work, hardly worth consideration. No glue enjoys a higher reputation than Scotch, and where it can be obtained no trouble need be taken with other sorts. Some of the French glues are also very good, and perhaps their light colour and clearness may be tempting. It must not, however, be forgotten that, however useful a light-coloured glue may be for some purposes, it is often less tenacious than the darker kinds. The best glue is of a clear, reddish-brown colour, very perceptible when looked through, and free from streakiness and muddiness. If on holding it up to the light it is cloudy instead of being clear, depend on it that the glue is not of the best. The dark, opaque-looking glue which one often sees in London shops, offered as "town made," at ridiculously low prices, cannot be recommended. When broken, glue should present a clean, sharp fracture, but this is a test which can hardly be applied beforehand when purchasing only a small quantity, and, with other indications which to an expert may mean a good deal, can hardly be of much use to the novice. He must be content to be guided to a great extent by the general appearance of the cakes, and a little comparison of different qualities which may come under his notice will soon enable him to distinguish them.

We may now consider how to prepare, or as it is generally called "make," the glue. It is presumed that the ordinary form of glue-pot, consisting of an outer pot to hold water only, and an inner one for the glue, is too well known to need any detailed description. In the absence of one of these, the glue may be made in an empty jam-jar, or anything of that kind; but on no account must the vessel containing the glue come in direct contact with the fire. If it does there is almost a certainty that the contents will be burnt, a risk that may be avoided by using a small saucepan as a substitute for the outer glue-pot. For ordinary domestic use a small glue-pot is preferable to a large one, or, at any rate, only a small quantity should be mixed at a time. The reason for this is that every time glue is melted it loses in strength, so that no more should be made at once than is likely to be soon consumed. However good the glue may be originally, by the time it has been several times melted it will have lost much of its virtue, and, so far as strength is concerned, be very inferior. To make the glue, break up a sufficient quantity into small pieces which will go conveniently to the bottom of the pot. It is not necessary to powder up the glue. Cover the pieces with cold water and let the whole stand for some hours. Do not use hot water in order to hasten the solution. Under the influence of the water the glue will swell and become soft; but please note that glue should not dissolve. It merely swells up and becomes soft. The time required cannot be exactly stated, but if left in soak overnight it should be fit for use next morning. It may here be mentioned that one rough-and-ready way of recognising the quality of glue is by noticing the quantity of water it absorbs. The more it takes (within bounds, of course—broadly speaking), the better it is; but it is impossible to specify this test more exactly, and the hint is thrown out to direct the novice's own observation.

When the glue is soft enough, it only remains to dissolve it by heat, the outer pot being partially filled with water. In a short time the glue will have completely melted, and may possibly require the addition of some more water; for it is a great and common mistake to suppose that the thicker the glue the better it holds. On the other hand, it does not do to have it too thin and watery. It ought to be thin enough, when quite hot, to run freely from a stick or brush dipped into it, much as ordinary paint or oil would, and thick enough when cold to form a tolerably firm jelly. Unless it gelatinises when cold, it is too thin to be of use; and, as it is better for general work to have it too thick than too thin, water should only be added cautiously. A little from the outer pot can be poured in among the glue whenever evaporation may

render this necessary. When the glue is hot, a very objectionable smell may sometimes arise from it; but this smell must not be mistaken for the ordinary odour of good glue, which, though very characteristic, can hardly be considered objectionable. We wish more particularly to direct attention to the peculiarly offensive odour of glue which has either been made from rotten material or has become putrid. Whichever it is, the glue must be condemned as being unmistakably bad, and in such cases the discovery cannot always be made till heat has been applied. As good glue also has an unpleasant odour when burning, the novice should learn to recognise this, though, indeed, one can hardly be confused with the other.

Enough has now been probably said about the preparation of glue, though it may be well to add that if the contents be left for any considerable time the moisture evaporates, till a solid mass remains, as hard as the original cake.

We now come to another part of the subject, viz., the application of the glue; for, however carefully prepared and good this may be, the full benefit cannot be derived without care and some knowledge of the way in which it should be used. It may seem an easy matter to dab some glue on two pieces of wood and stick them together; but how is it one so often finds that breaks which have been mended at home are not effective? The two pieces joined together come asunder again on the slightest provocation, just exactly at the joint. Now, unless the wood be of a very tough and strong kind, the glued joint ought to be at least as strong as the wood itself, i.e., the glue holds on so tenaciously that, as in the case of two boards of pine connected by their edges, an attempt to pull them apart would probably result in the split being in the solid wood in the direction of the grain, and not at the glued junction.

This leads us to say that two pieces of end-grain wood cannot be securely fastened with glue; hence the futility of any attempt to join such a thing as a broken chair-leg with it. Of course we refer merely to those instances where part of the leg is snapped off. If merely a piece is split away it is quite a different matter, for in this we have the fracture concurrent with the grain of the wood, and consequently there is a legitimate opportunity for glue. Whether for an original joint or a repair, however, the same general principles must be regarded where strength is an object. To begin with, the two pieces must fit each other exactly. Glue will fill up any interstices there may be through clumsy fitting, but, though it fills them, it will not make by any means a strong joint. The two pieces to be united must be warmed where the glue is to be applied, so that the latter may not be chilled when it is rubbed on. It is possible to have the wood too hot; but this is a mishap which is not so likely to occur as having it too cold, as, by the time the wood is hot enough to injure the glue, it is almost burning, or at any rate scorching. So, perhaps, all that it can be necessary to say is this: Have the wood hot, but do not burn it. The glue itself being thoroughly hot, a small quantity must be rubbed on one of the edges, and not only on but well into the wood. As quickly as possible the other edge, which is to be also warm, must be applied. When the edges are straight, instead of merely pressing them together it is better to slide and work them slightly against each other, as any air bubbles which may have formed between them are more certainly got rid of, as well as superfluous glue being squeezed out. This cannot, however, be managed when a break is being repaired, and direct pressure alone must do. As much glue as possible should be pressed out; the closer the two pieces are brought together—or, in other words, the less glue there is between them—the better will be the result. No greater mistake can exist than to suppose a quantity of glue is required in the joint; nor is it enough just to press the two pieces together. The pressure must be maintained till the glue has thoroughly set or hardened, and the time for it to do so will depend on such a variety of circumstances that no details can be given beyond saying that a small joint left to dry in a warm room should be sufficiently firm in a few hours.

It may occasionally happen that a joint which has been glued once has come apart and requires rejointing. Before this can be done the old glue must be thoroughly removed, and the same may be said even if the glue has only recently been put on, as, for example, when, through an error in supposing the joint to be firm before it is so, the pieces have been pulled apart. The glue which has exuded may easily be removed by a chisel when it has partially congealed, or even when it has become quite hard.

As a rule, good glue is sufficiently tenacious for ordinary woodwork, but it may be strengthened by the admixture of a small quantity of brickdust or plaster of Paris. It is as well, however, to avoid adding anything unless absolutely necessary; and as for the various recipes—of which many have been published—for keeping glue always ready for use in a liquid state, it may be said that none of them can be recommended where a strong joint is a primary consideration. Glue certainly may be kept liquid and not require heat to reduce to a usable consistency; but convenience is attained at the expense of strength whenever this is the case.

This remark does not apply to such glues as Le Page's, which have deservedly become popular for domestic use, and no article on glue for household purposes could be complete without some reference to them. That best known is Le Page's, which is sold in bottles and tins in so many shops that it can hardly be looked on now as a novelty. Except in very cold weather, no preparation is necessary to render it usable, and, even in extreme cases, putting the tin or bottle in warm water will effect all that is required. When bought in small quantities, Le Page's glue is more expensive, bulk for bulk, than the ordinary kind; but remembering how much of this latter must necessarily be wasted where only a little is used, and that there is practically no waste with the former, there is not so much difference from an economical standpoint as appears at first. As with ordinary glue, the great secret of success is not to leave too much in the joint.

Perhaps it may be news to some that there is no better brush for gluing purposes than a piece of common cane, hammered at one end till the fibres are loosened and form a stiff brush. Before hammering, the hard outer coating of the cane should be removed.

AN EASILY-MADE FOUR O'CLOCK TEA-TABLE.

THIS table is very simple in construction—there are no mortice joints, dovetails, or lathe work, &c., which so often deter amateurs from attempting to make such articles of furniture. The only tools required are of the simplest kind, and the pattern is not so common as the stereotyped "gipsy," with its three turned legs meeting in a ball. The top may be either round, or shaped as indicated by the dotted lines (Fig. 46), can be made of either pine or deal, and is 24in. in diameter.

Procure a piece of yellow pine, if possible 24in. square by 1in. thick, or if so wide a board is not procurable, join narrower ones together, truing the edges and joining with glue, then clean both sides quite smooth, finishing with sand-paper across the grain. Now describe a circle of 12in. radius, and cut out with a lock saw, leaving a slight margin to allow for finishing with a sharp spoke-shave. Next draw two diameters at right angles, and then lines parallel to these, and ½in. to each side of them, and lastly describe a circle of 9in. radius.

We now have the position of the legs indicated by the portions shaded by diagonal lines (Fig. 46).

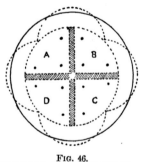

FIG. 46.
UNDER-SIDE OF TOP OF TEA-TABLE.

Fig. 47 shows the shape and dimensions of the legs. They should be of well-seasoned beech, and fully 1in. thick

when finished; therefore select timber a shade stouter to allow for cleaning. Having cut out and finished one leg, mark and cut the other three by it, and see that all four are exactly the same in size and shape. The front edges of the

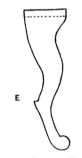

FIG. 47.—LEG OF TABLE.

legs should be slightly chamfered. Now, upon a piece of board 1in. thick (any kind of timber will do) describe a circle of 9in. radius, cut it out with a fine saw, and smooth the edges. Draw two diameters at right angles, and then cut the circles into four segments. Plane the straight sides true, and glue securely in the positions indicated (A, B, C, D, Fig. 46). See that the spaces shaded diagonally are exactly of the width of the thickness of the top ends of the finished legs. When these segments are glued, secure further with three 1¾in. screws through each, as suggested in Fig. 47.

Now fit the legs into their places, temporarily, set the table standing, and measure the distance between the legs at the point E. Supposing this to be 10in., next cut out a circular piece of ¾in. pine, or beech, &c., 10in. in diameter, which is to rest in the notches (E) in the four legs, forming a shelf, and also serving to give rigidity to the whole.

Now glue the legs into their places, fixing the circular shelf in the notches, and securing by one 1½in. screw at each notch, in an oblique direction, counter-sinking for the screw-heads, which can be covered with putty before painting.

For further security two 2in. screws may be inserted through the table from the top into each leg; but this will not be absolutely necessary.

The legs should now be painted in three coats of Aspinall's enamel, &c.; the top covered in plush, or otherwise ornamented according to taste.

AN EASILY-MADE DWARF BOOKCASE.

THE dwarf bookcase, (Fig. 48), the subject of the present article, has all difficult joinery work dispensed with, although no one can say that its looks betray its peculiar construction, or that it is not strong. Simplicity of construc-

FIG. 48.—DWARF BOOKCASE.

tion, resemblance to work of the ortho-dox kind, and substantiality, are the three requisites.

The dimensions given are those of the example, but of course they can be altered to suit requirements. The chief measurements are as follow: Height to top, not including the back-guard, 3ft. 10in ; depth from back to front, on end of top, 1ft. 1in. ; and width over all in front, 4ft. The shelves are movable, so that they can be adapted to various sized books. As for the wood, it is pine —part of an old packing-case. The one the writer used had contained either a Continental piano or an American organ, and the boards of which it was formed were grooved and tongued, besides being fairly smooth ; the boards are a nice handy size, and part of them being fitted to each other saves a good deal of labour in jointing up, as, the grooving and tongueing being done by machinery, they all fit each other.

Where widths are mentioned, it will be understood that if there are no boards al-ready wide enough, they are fastened together to make up the required size and the sur-plus ripped off with the saw. Of course the pieces were glued together and tightly cramped up.

The sizes of the principal parts are as follow: Top 4ft. by 1ft. 1in., ends 3ft. 9in. by 1ft., and bottom 3ft. 8in. by 11in.

The top is nailed on to the ends, flush at the back, but overhanging 1in. in front and 1in. at each end. The bottom-board is also to be nailed to the ends, but within them, and 5in. from the ground, the edges being flush in front, consequently the bottom-board does not come within 1in. of the back edges. In fastening these pieces together, careful use should be made of the square to insure correctness, and the same caution should apply throughout. This presupposes that all the boards are cut squarely at the ends, for unless they are it will be impossible for the fitting to be true.

The principal parts of the case are now together, and Fig. 49 represents the work at this stage. A piece of wood 3in. wide, and the same length as the bottom-board is wanted, to serve as the frieze,

FIG. 49.—SKELETON OF BOOKCASE.

and is nailed within the ends, just under the top, through which a few nails may with advantage be driven into it. Where it shoulders against the ends see that it is perfectly level with them; a couple of nails at each end will be sufficient to hold it. The frieze should be placed with an edge, not a side, against the top—*i.e.*, looked at from the front its width, not its thickness, is seen. Now take two pieces 1in. square, and long enough to reach from the under edge of the frieze to, or near to, the ground. These are to thicken up the front edges of the ends, and also to hide the racks on which the movable shelves are to be supported. Before they can be fixed, a piece 1in. square must be cut out of the front corners of the bottom-board. This may be easily managed by sawing across to the depth of 1in. from the front, and chopping the small piece away with the chisel. The pilasters are then to be glued and nailed along the inner edge of the ends. If they do not reach quite to the ground it will not matter, as a plinth will cover this part; but they should fit closely up to the frieze, and the fronts be level with those of the ends. The result will be that the ends, looked at from the front, appear 2in. thick.

The back may next be prepared and fixed, or left till later on. The boards forming it should run from top to bottom, not across from end to end. Before nailing on the back, a rail similar to, but not necessarily so wide as, the frieze in front should be fastened under the top, and instead of being flush it should be set back 1in. The back-boards, with their upper ends against the top, may then be nailed to this piece. The bottom-board, it will be remembered, is also 1in. from the back of the ends, so that the pieces enclosing the back rest against and are nailed to it. They may go to the ground, but as they are not seen, and as no object is served by their doing so, it will be quite sufficient if they are long enough to go behind the bottom-board.

With the exception of the shelves, the bookcase is now to all intents made, and those who do not want to expend more labour may nail these in. As it stands, however, the bookcase will be only a crude affair, and the ornamental details may be attended to.

Underneath the overhanging top a moulding is fixed of the section and sizes given in Fig. 50. These and similar machine-run mouldings in pine are to be bought at a very cheap rate in London and other large towns. They are sold in lengths of about 12ft., but at a rate per 100ft. from about 1d. per foot. At the front corners the mouldings will

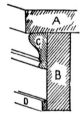

FIG. 50.—FRIEZE AND MOULDING.

have to be mitred. Those who have not the tools or skill will have no difficulty in getting some friendly joiner to cut the mitres in a few minutes accurately, if the length of the front of the bookcase is exactly given. As the ends of the mouldings are cut-off level with the back it will not be necessary to give the length of the end pieces, for, provided they are long enough, nothing more will be required. Any excess can be sawn off after they are fixed. The moulding should be fastened with glue and a few brads to the bookcase; but when driving these in be very careful not to bruise the moulding with the hammer. If, however, any accidents of this description occur, bruises may be removed by wetting them and allowing the moisture to soak in. If the fibre is not too much fractured the bruised parts will, under the swelling influence of the damp, soon disappear.

Along the lower edge of the frieze, a strip of wood about ⅜in. wide by ¼in. thick should be glued and bradded. This is shown by D on Fig. 50, where A represents the top, B the frieze, and C the moulding.

At the bottom of the bookcase the plinth has still to be fixed. It consists merely of three pieces—one in front and one across each end—5in. wide, or exactly the measurement from the ground to the upper surface of the bottom-board. These three pieces may be mitred at the corners, but a less troublesome way is simply to let the front piece overlap the end pieces. The top front edge of the plinth should be bevelled (Fig. 51) before fastening it to the job with a few nails driven through into the ends and into the bottom-board. The corners of the

FIG. 51.—PLINTH.

overlapping front portion of the plinth may with a chisel easily be cut away to correspond with the bevel along the ends.

To form the racks, cut four pieces ½in. thick from a board, and after planing them up, fasten them together with two or three screws, so that they form a solid block 2in. wide by 1in. thick. Now, at intervals of about 1in., with the aid of the square, mark lines across the four pieces on one side, taking care that the lines are equi-distant from

each other. When this has been done, saw down on each line to the depth of say ⅜in. Then, with a chisel, cut out slantingly from the top edge of one cut down to the bottom of the next, and when all have been treated in the same fashion we have the four racks, one for each corner of the bookcase. Each piece should be long enough to reach from near the top to the bottom-board. They are then either glued or bradded to the ends of the bookcase—inside, of course—great care being used to see that each pair is perfectly parallel, or the rails on which the shelves rest will not fit properly. The rails are the same thickness as the racks, and have their ends shaped to fit into them. Fig. 52 shows the racks with movable rail. A

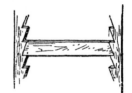

FIG. 52.—RACK AND RAIL FOR SHELVES.

pair of rails will be necessary for each shelf, and it may save some subsequent annoyance if, when making the bookcase, the precaution is taken of making them all fit anywhere. As the weight of the shelves keeps the rails firmly in their places it is not necessary that the latter should fit tightly; indeed, it is better almost that they should not, as they are then more easily removed when desired. The shelves themselves require no special mention, all that is requisite being that they should be long enough to rest on the rails, and to allow them to do this a recess must be cut off in each corner for the space occupied by the racks.

The leather edging may be used or not. It certainly gives an appearance of finish at a very trifling cost of money or labour. If the edges of the shelves are nice and clean, the edging will look better if fastened below, as shown in Fig. 48, otherwise by gluing them to the front of the shelves any unsightly appearance there may be covered. To fasten it under the shelves glue the leather on to slips of wood, and either nail or glue these to the shelves. The appearance will be better if these overhang a little (*i.e.*, the leather should be set back from ½in. to ¼in.). Care must be taken that the top edge of the leather does not project above the wooden slip, or this will not fit closely to the shelves. Like the leather, the back-guard may

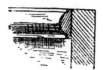

FIG. 53.—MOULDING ON BACK-GUARD.

be omitted, but besides being slightly ornamental, it is useful to prevent articles placed on the top of the bookcase

from falling down behind. As shown, it is simply a piece of board, 6in. wide, nailed on behind the top, slightly rounded at the ends, and with a piece of moulding glued on along the front with the ends mitred on. The moulding is shown in section in Fig. 53, but any of similar character, or even a piece of the same as used on the frieze, would do very well.

The pilasters in Fig. 48 show a couple of beads. These are purely ornamental, and are easily worked by a home-made scratch or router.

The bookcase when made can be coated with an enamel paint after the nails have been well punched in, the holes filled with putty, and the knots rendered innocuous to the paint by going over them with "knotting," a preparation sold at paint shops for the purpose. The result is better than would seem possible.

A LAVATORY GLASS.

A LAVATORY glass, which is a combination of mirror, toilet-table, and towel-rail, small in size but very useful, is shown by Figs. 54 and 55, the former giving the front, the latter the end elevation. It consists of a frame containing the glass, a small drawer below, and lower still a rail for the towel. It will be described in its entirety; but for those who wish to make something even simpler, a few suggestions may be made, all of which, on the principle of the greater including the less, there will be no difficulty in putting into practical effect if the construction of the whole is understood. First of all, the glass may be in a frame only without the additions of towel-rail and drawer. Next, there may be a shelf, but no drawer, or instead of a drawer under the shelf the space may be used as a cupboard, or there may be two drawers. Whichever

arrangement may be considered most convenient, the principles of construction are almost the same. The sizes must be left for individual taste or requirements to decide, but as a guide for those who have no very definite ideas the following directions are given as likely to be generally useful: The size of the plate—*i.e.*, the silvered glass—15in. long by 12in. wide; the width of the shelf, from back to front, 8in.; the depth of the drawer, 5in. (this latter measurement being given as outside the front, so that the actual depth inside will be from ½in. to ¾in. less), and a towel-rail, some 4in. or 5in. below, will give a convenient lavatory glass, neither so large as to be cumbersome nor yet so small as to be merely an apology for something better. The wood to be used is a matter of choice, but should be clean and free from knots. Whatever the wood, let it be thoroughly well seasoned; and with this caution let us proceed to practical details.

To begin with, I strongly recommend the maker to prepare a full-sized work-

FIG. 54.—FRONT VIEW OF LAVATORY GLASS.

ing drawing, not necessarily showing all the constructive details, but "setting out" the work to the actual sizes it is to be made up. This will save a great deal of uncertainty, not to say waste of material, in cutting the various parts, and though it may seem a needless trouble it will save time in the end. The setting-out may be easily accomplished by anyone who can use a pair of compasses in conjunction with a rule to get sizes, and rule the lines according to the dimensions so obtained. Fine drawing is not required, the principal point being correctness and distinctness. Beyond that two drawings will be required, one of the end and the other of the front, very similar to Figs. 54 and 55, nothing need be said about this preliminary part of the work. The frame

will be of 1¼in. square stuff, and assuming that the sizes already given are to be worked to, we shall require the following approximate lengths to form it, allowing a little for squaring off ends and jointing the cross-rails: Two pieces 2ft. 6in. long for the sides, and three pieces 12in. long. Of the latter one is for the top part of the frame, another for the bottom, and the third is merely a stay to which the bottom of the drawer-box is to be screwed. These three may be fastened to the uprights by mortice and tenon, by dowel joints, or by the less workmanlike but, to unpractised hands, easier way of halving them in from behind, when a screw nail at each joint will be required to bind the whole securely together. As the last plan is not so common as the former two, and is only likely to be adopted by beginners, it may be useful to explain it

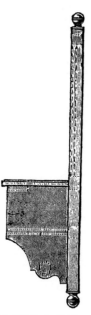

FIG. 55.—END VIEW OF LAVATORY GLASS

somewhat in detail; though, as will be seen from Fig. 56, it is so simple that it can hardly be necessary to do more than say that the ends of the cross-pieces are

D 2

cut down from the front to a depth of, say, half the thickness of the wood, and a recess made in the back of the upright to such a depth that when the other is placed in it the front surfaces of both are

FIG. 56.—METHOD OF HALVING-IN CROSS-RAILS.

on the same level. It will be noticed that the halved piece of the cross-rail does not run quite through the upright, as, were it to do so, the joint, viewed from the side, would be unsightly, and at once proclaim the unusual construction.

When the five rails mentioned, viz., the two long uprights and the three shorter connecting them, are put together, we

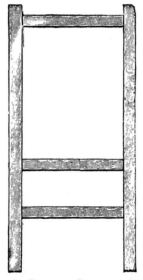

FIG. 57.—FRAME OF LAVATORY GLASS.

have the frame as shown in Fig. 57, the large open space between the top and middle rail being for the glass. The

next space downwards is covered by the drawer-box, of which the top is screwed to the middle rail and the bottom to the lowest one. A rabbet will have to be cut in the frame for the glass, and, of course, this will have to be done before the parts are permanently fastened up. The rabbet should come to within about ¼in. of the front, but its width is unimportant, provided it is enough to hide the rough edges of the glass. In the long pieces the rabbet may be cut just the required length, but if worked with the plane it will be found much simpler to run it from end to end and fill up the space where the uprights show square (i.e., the inch or so above the top rail and the parts below the drawer-box). This, as well as the following for forming a rabbet by "facing," is a method very commonly adopted, and, except for the theoretical objection that the work is not cut from the solid, nothing can be urged against it. Whether a legitimate form of construction or no—and there is no occasion to discuss the matter here—there can be no doubt that it will be welcomed by amateurs who cannot work a rabbet-plane, which is not such a universally-employed tool as some others, though a very useful one. There are, however, generally more ways than one of doing any operation in joinery, and this is one of the "other ways" of

FIG. 58.—SECTION OF RAIL, FACED UP.

forming a rabbet. Perhaps it is best explained by reference to Fig. 58, which shows in section the rail "faced up." The "facing" is 1¼in. wide, and about ¼in. thick. It is, as its name implies, stuck on the face side of the column, which, in this instance, is 1in. square, instead of being 1¼in. as named for the uprights and rails from which the rabbet was to be cut. One edge of the facing being flush with the outside of the frame, whether rails or uprights, the other must necessarily overlap, and so form the rabbet. The facing is glued on, and, provided the wood has been truly

planed up, the joint will be so close that
the principal indication of it will be the
variation in grain of the two pieces—
indeed, unless closely looked for, it will
not be perceptible. If the frame is to
be formed in this way, the backing—
i.e., the 1in. squares—should be made
up, the facings to the end pieces glued
on, and those for the cross-rails measured
and fixed accordingly. If the facing
pieces on the outside ends overlap, the
surplus can easily be removed by running a plane along. Of course with
"facing" there will be gaps left similar
to those left by the rabbet-plane when
run the whole length of the wood, and
it can hardly be necessary to say that
they should be filled up as already indicated.

The frame may now be considered
ready for the drawer-box and fittings.
The top, as already stated, will be about
8in. wide, and the length should be
exactly the same as the extreme width
of the frame, or if there is any difference
it should be a trifle less, for if it overlaps the frame it will be an unsightly
piece of work. The brackets or side
pieces enclosing the drawer may be attached to the top in two ways, or rather
there will be no occasion to mention
more than two here, as others are merely
modifications of them. Let us take as
the first, and perhaps the better of the
two, the one shown on Fig. 54. It will
be observed there is between the top of
the box and the drawer-front a straight
piece of wood, which, as seen by the
dotted lines immediately under the top
in Fig. 55, is in front flush with the edges
of the bracket sides, and is a few inches
in width. This piece serves to bind the
ends together as well as affording a
support to the top, which is screwed to
it from underneath. Before going any
further, let me say that ½in. wood is
thick enough for the ends and top, but
¾in. will be better, as this will not finish
more than ½in. at the most. The ends
of the box should be fitted to the frame
with an equal overlap on each side, so
that to get the exact length of the piece
of wood under consideration it is only
necessary to measure across the frame
from centre to centre of the uprights.
It is then dovetailed into the upper edge
of the ends as shown in Fig. 59, and when
all is ready a couple of screws may be
driven through it from below into the

top; but before this can done the rail
itself should be glued in, and if a nail or
two are used at each end, so much the
better. As they will not have much

FIG. 59.—DOVETAIL JOINT.

hold if driven in straight, insert them
in a slanting direction, when it will require considerable force, even if the glue
perishes, to pull the top piece up. As
will be understood, the width of this
rail or stay is unimportant, but 3in. or
4in. will be convenient, for, if much less,
slides will have to be fixed for the upper
edges of the drawer sides to bear against.
It may either be of pine faced up with
the principal wood of the article, or
solid; in either case the same thickness as the top will do very well.

The alternate method of fastening the
top on is still simpler, consisting merely
in fixing it to the sides with dowels;
but unless the holes are accurately
bored in the top to correspond with the
dowels in the end, it is not likely to be
a success. Still, with a very moderate
amount of care and skill this may easily
be managed. It may not be so important to draw attention to this as to the
desirability of not boring through the
top, a mishap that might easily occur.
To prevent it, as soon as the pin of the
centre-bit (which will be the best tool
to use for boring) makes its appearance
through the wood, do not drive it any
farther. The dowels, of course, must be
glued in both to the top and ends; but
do not fill up the holes with glue—a
very common mistake. Also be careful
that the dowels just fill the holes, cutting them as nearly as possible to the
exact length.

Now for the bottom of the drawer-
box. This may be either grooved in or
dowelled in to the ends, it really does
not matter which. If the latter, all
that has been said about this method of
fastening the top applies equally to it;
but a few remarks about the other may

not be amiss. The ends of the bottom may be either let into a dovetail groove, in which case they must be shaped accordingly, or be fitted into a plain trench or channel cut across the ends for them. The former is a somewhat difficult job, and in a small piece of work like the present there is no real necessity for it, though, if it can be managed, there is no objection beyond the extra amount of work. A plain groove, however, ought to do just as well, and all that it can be necessary to say is that it need not be more than $\frac{1}{4}$in. deep, just the width of the thickness of the bottom, which should fit tightly into it, and that it should stop a little way ($\frac{1}{2}$in. or so) from the front. A small piece will, of course, have to be cut away at the corresponding corners of the bottom piece to allow the ends to fit into the grooves.

The towel-rail has still to be considered. It may consist of a round piece of wood about 1in. in diameter, or if there is a difficulty in making a presentable round rail, it may be octagonal in section, and formed by planing the edges off a 1in. square stick. To support between the ends it will be only necessary to bore holes part of the way through these with, say a $\frac{1}{2}$in. centre-bit, and cut the end of the rail to fit into them.

When all these parts have been fitted to each other, they may be permanently fastened together, and the manner of doing so is so clear after what has been said, that no further directions can be necessary. Let us suppose, therefore, that the box part has been made up and fastened to the frame with a few screws driven in from behind, say two into each side piece, three into the top (one of them through the cross rail, and one through each upright), and the same into the bottom board, in all eight screws, which should be quite sufficient to make the whole perfectly rigid and strong.

The drawer has still to be made. As there is nothing special about it no minute details need be given here; but it may be well to remind those who are not quite clear how to go about it, that the first thing is to fit the drawer-front to the space it is to occupy, and to see that it fits tightly, for any "easing" can be done after the drawer is made.

The next thing is to prepare the drawer sides, letting them also fit tightly. The back and bottom being prepared, all the parts can be fitted together in the usual manner.

With the exception of the finishing touches, the work may now be considered complete so far as the joinery is concerned, if we except the small knobs, which are shown at the ends of the uprights of the frame in Figs. 54 and 55. These finials should be turned, with a pin to fit into holes bored for them. Should it not be convenient to make them with the wooden pin, a double-ended screw may be used instead, the principal objection to this being that unless care is taken the knobs are liable to be split when making the screw-holes in them. A small brass handle should be fitted to the drawer-front, the screws used to attach it being also brass, and preferably with rounded heads. This may seem a small matter to direct notice to, but it is really by attention to such apparently insignificant details that a great deal of the finish of any article of furniture is determined, and it must be remembered that it is quite as important to have a well-finished thing as to have it soundly made. Under the same category comes the polishing, and let me say again that polishing, not varnishing, is meant. Do not spoil the work by a coating of varnish, which may give a gloss to—but cannot enhance—the natural beauty of the wood, as properly executed French-polishing ought to. Nothing has been said about fitting the glass in, as unless it is to have bevelled edges no difficulty can occur, even to the most inexperienced, all that is necessary being that the plate should fill the rabbet when a thin pine backing let down on it and bradded in will obviate any necessity for blocking the plate in. If the plate is bevelled this will have to be done, but to explain the method would be merely to recapitulate the instructions recently given in these columns in the articles on "Furniture Glass and Glazing," to which those who are not well up to the work will do well to refer. Fig. 54 shows the rails, &c., with beads, but these being merely decorative may be omitted if preferred without in any way altering the construction. That they do however relieve the surfaces from absolute plainness and give a character to the whole is undoubted, so those who can will do well to work them on. It must, however, not be forgotten that a plain surface is not in itself ugly; its beauty may be of a negative kind; but the same cannot be said of badly finished or executed work that is added for the sake of effect. Therefore my advice is not to attempt "beading," unless it can be done properly. Let the first attempts be on waste material, or, at any rate, on work where comparative failure will not be so conspicuous as on the lavatory glass, which one may almost say, is "all front," so that it is out of the question that any superficial defects can escape notice.

The directions given in this article are more by way of suggestion than definite as to mode of manipulation, but it is to be hoped that they will be none the less serviceable, not only to the skilled amateur, but also ¦to the novice who does not care to be continually in leading strings.

A KITCHEN TABLE.

THE drawings (Figs. 60 and 61) show two different patterns of kitchen tables—Fig. 61 being perhaps the better pattern, as it takes up less room when the flap is down. The frame is made

FIG. 60.—KITCHEN TABLE.

exactly the same in both cases, only it is 9in. less in width in Fig. 61

The table is made in common deal, the top and sides of frame being 1in. thick. The legs, 2¼in. by 2¼in., made out of deal quartering, are 2ft. 7in. long, with a slight taper, starting 6in. from the top, making them 2in. by 2in. at bottom. This is just enough to take off a heavy appearance, without weakening the table, for it is necessary that a kitchen table should be very firm, capable of standing hard usage.

In Fig. 60 the top measures 4ft. 3in. long, 3ft. wide, and should overlap the frame about 1in. on all four sides. The frame is shown in Fig. 62. The front rails, sides, back, and division between the drawers are of 1in. stuff; and the front rails (A and B) 4in. wide, and 3ft. 8in. long when fitted. The top rail is let into the legs by means of a tongue, 1½in. long, each end, which must be allowed for when cutting the rail out. The rail (B) is mortised into the legs, allowing a space of 4in. between the rails for the drawers. The tenons of this rail are 1¼in. long, 1½in. wide, commencing ½in. from the outside edge, so that there will be a ½in. margin to the mortice in the legs. The side-pieces measure 2ft. 5in. long when fitted, and 5in. wide. These, too, have tenons 1½in. long each end, so must be cut at first 2ft. 7½in. long. The tenons are 4in. wide, starting 1in. from the top. The mortices in the legs cannot be more than 1¼in. deep, as those for the side-pieces will meet those for the front rail and back at right angles. It is as well to make the mortices in the legs first before tapering them, as they will then lie flat on the

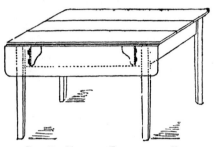

FIG. 61.—KITCHEN TABLE, WITH FLAP.

bench. The division between the drawers is 5in. wide, and is screwed to the top and bottom front rails, and either mortised or screwed to the back. The part that fits between the front rails will necessarily be 1in. less in width. The runners for drawers can be cut out of ¾in. wood—the side ones 4in. wide, the middle one 6in. ; these are screwed on to the sides and division from underneath. This will complete the frame.

The top is made of four 9in. deals joined together; all the inner sides must be grooved ¼in. deep, ¼in. wide, and slips of wood inserted, the grain of which should run at right angles to the boards themselves. As in every case of making joins, the edges must be planed perfectly true. Many amateurs possess a proper shooting-board, by which the edges can be shot perfectly true. When the grooves are made and the slips or tongues inserted, the boards must be glued together and put in clamps till the glue is hard ; then finish off the top with a smoothing-plane, and round the front corners of all four. The grooving and tongueing is necessary for two reasons : First, as kitchen-tables are usually left plain—that is to say, not painted or varnished—they are frequently cleaned, and unless the joints are tongued the

water will penetrate in time into the drawers ; and, secondly, the tongues prevent the top warping after being washed. The top is then nailed or screwed on to the frame.

The drawers will measure 1ft. 7in. long, 4in. deep, and about 1ft. 3in. back to front. The fronts should be of 1in. stuff, the sides and backs of ½in., and the bottoms same thickness. The drawers are made the same as those described in the article on "A Washstand." The handles or knobs, in wood or china, can be obtained at any ironmonger's, and should be 2in in diameter.

In making a table as shown in Fig. 61, the flap will of course be at the back, and the fixed top will consist of three 9in. deals tongued and glued together, overlapping the frame about 1in. at sides and front, but flush with the legs at back. The frame mentioned is made the same as in Fig. 60, but will be about 9in. less in width. Cut out two wooden brackets similar in shape, as shown in Fig. 61, 6in. long, 7in. wide at the top. These are secured to the back with iron

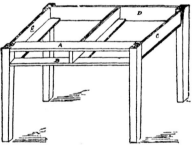

FIG. 62.—FRAME OF TABLE.

or brass hinges (2in. cast-iron hinges), and they must be placed so that the tops will lie in just under the fixed top of the table when not in use. The flap is a 9in. board secured to the table with three 2½in. cast-iron hinges, which must be let into the edge of the flap, so that it will fit close to the fixed top when in use.

A PLATE-RACK.

A PLATE-RACK, if not actually a necessity, is certainly so very useful an addition in the domestic department of a house, that it should recommend itself to the amateur to make one, if he does not possess such a thing among his household furniture. True, it can be purchased at a merely nominal price, but it is well worthy of the amateur's attention.

The drawing gives a general idea of a plate-rack, and shows the rails on one side, though of course both sides are made the same. The rack can be made in various sizes; but the one shown is a fair average, and has four divisions for dishes and twenty 2ft. divisions for

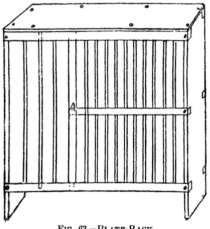

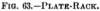

Fig. 63.—PLATE-RACK.

plates. It measures 10in. long, 7in. wide, and 2ft. 6in. high at the ends. The top and bottom rails are 1½in. wide, 1in. thick (or ¾in. when planed will do), the end-boards ½in. or ⅝in. thick, the rails being let into the boards, as shown in the drawing, and screwed in, allowing a space of 2ft. between top and bottom rails. The uprights (A) are 1in. square

and mortised into the rails. The middle rails of the same thickness are mortised into the uprights (A), and let into the end-board in the same way as the other rails. The uprights (A) are to be fixed 10in. from the left-hand side.

Before fixing the rails finally to the end-boards, the holes must be bored for the rods. To facilitate this, clamp the four rails side by side—i.e., the two top and two bottom—and, starting from the left hand, mark out where the holes are to be for the dishes. These will be 2½in. from centre to centre (the whole division being 10in.), and for the plates 2¼in. from centre to centre. Drill the holes with a brace and a ½in. twisted bit; the top rails to be drilled right through, the bottom rails half through; the middle rails are also drilled through. These should be marked out first from one of the other rails, so that the holes should be exactly in a line with the others. The rods to be cut out ½in. square, and the edges planed off; it not being necessary that they should be perfectly round, but only sufficient for them to slip through the holes. The rods for the plate division can be ⅜in. if preferred, in which case the holes will be drilled the same size. This being done, screw the rails to the boards, and put in the rods from the top. In making the rods, let them be fully long, for when put into position they can be cut off flush with the top. Finally, screw a ½in. board on to the top, which not only makes the rack firm, but serves as a protection from dust. There is no need to fix the rods otherwise than by fitting them tightly into the holes, and on no account should glue be used. The bottom is, of course, left open, to allow of the plates and dishes draining; and the rack should be fixed, if possible, over the scullery sink by means of holdfasts driven in the wall and screwed to the side-boards.

HOUSE-STEPS.

HOUSE-STEPS are practically a necessity in every home, and are easy to make, as the following description will show:

As to size, a six-tread is the most useful for ordinary houses, that is, where the rooms are of ordinary pitch. In that shown by Fig. 64, the height is 5ft. 2in. without the top, the width of sides 4½in., the ends being cut at an angle of 22° 30′ (as shown in A B, Fig. 65). The thickness

edges to be planed at the same angle as the sides are cut, the extra ½in. in width allowing for this; the fronts also project

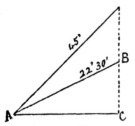

FIG. 65.—ANGLE OF ENDS OF SIDES.

slightly beyond the sides, the front corners of the treads being bevelled off to meet the sides.

Before proceeding further, it may be noticed that the steps as drawn are

FIG. 64.—HOUSE-STEPS.

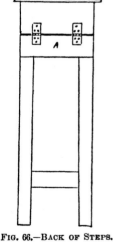

FIG. 66.—BACK OF STEPS.

of the wood to be used throughout can be either ¾in. or 1in.; but for indoor use ¾in. will be found strong enough. Run a bead on the front edges of the sides with a ¼in. plane, or leave them square. The treads are let into grooves cut in the sides parallel with the ends, at equal distances of about 10in., according to the thickness of wood used; the length of the five lower ones being 13½in. (which allows ¼in. each end for letting in the grooves), and the width 5in., the

straight—*i.e.*, the sides run parallel from top to bottom—whereas usually they taper slightly. The amateur can use his own discretion about this, and should he prefer them tapered, the second tread must be cut 12½in. long and the bottom one 14½in., the intermediate ones of course being in proportion. Straight steps will be the easier to make, and the length of tread which, when fitted, gives 13in., is sufficient to keep the steps firm.

To proceed: Having cut out the treads, glue them into the grooves, and screw or nail them from the outside. The top tread should be rather larger, projecting slightly on all four sides, the measurement being 15½in. by 6½in., with the front and back edges planed at the same angle as the others. To fix this, cut two grooves, ¼in. deep, into which the tops of the sides are glued, and screw in from the top. Underneath the top step at the back screw a piece of wood 6in. wide, to which the back is hinged. The back (Fig. 66) measures 4ft. 8in., the same width as the front, and of the same thickness. The top crosspiece, A, is 4in. wide, into which the uprights are mortised, they being 3in. wide; the lower rail (3½in. wide) is mortised into the uprights. The back is screwed on to the front with two iron backflap hinges, 2in. wide. Two stays of thin cord from back to front on each side will complete the steps.

If steps of more than six treads are made, the wood should not be less than 1in.

A BUTLER'S TRAY AND STAND.

A BUTLER'S tray, if bought at a shop, is generally of mahogany, but for the amateur it will look nearly as well made of deal, carefully stained and oiled (not varnished, in case of hot things being placed upon it). If mahogany is procurable by all means use it, though great care will have to be taken when planing it, as the grain is not at all regular, so that the smoothing-plane should be very sharp and finely set, for to tear the grain means a considerable amount of extra planing to make the surface smooth again; and also in mortising the stand together, the chisel must be in good order, otherwise the wood will split.

The following are the measurements for one made in ordinary deal, which has stood hard usage for three years and is now in every way as good as new. The amateur can of course alter the measurements to suit himself. Fig. 67 shows the tray and stand complete. The tray is 2ft. 8in. long, 1ft. 11in. wide, and 2¾in. high, not including the bottom, the front being ¾in. less in height, as shown.

The sides should be dovetailed together, and the bottom screwed on to them. The sides of bottom to be rounded and to project slightly beyond sides of tray. The width of tray will necessitate the bottom

FIG. 67—BUTLER'S TRAY AND STAND.

being of three pieces, which means two joins; these should be glued and clamped together till dry, the boards to be full long, and when joined cut to the exact size required. The two long sides of the tray have holes cut in them, as shown, for lifting. Under the tray screw two fillets, 1 in. by ¾in. from side to side: these not only tend to strengthen it, but prevent any slipping when on the stand; the positions of the fillets will therefore be inside the rails of the stand.

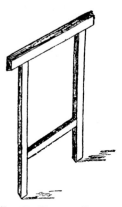

FIG. 68.—PART OF TRESTLE.

Fig. 68 shows part of the trestle, or stand: the height of each part is 3ft. extreme, the size of the wood used throughout being 1¾in. by 1in. The top rails are the same length as the width of the tray, viz., 1ft. 11in., and are planed at an angle on the top sides, so that when the trestle is extended for use the tray rests flat on them. Be careful to notice that the two uprights of one side of the trestle fit between the two uprights of the other side. For instance, the uprights of one rail are mortised into it, say 1¼in. from either end, so that the uprights of the other rail should be 3in. from either end. The cross-pieces should be about 8in. or 9in. from the bottom. The pivots on which the trestle swings are easily made by using two stout screws, 4in., No. 18. These are screwed in from the outsides, the thread of screw being in the inner uprights, the plain part of screw working freely in the outer ones, the heads countersunk flush with the surface; then file off any of the screw that may appear through. Two pieces of chair-webbing fixed to the top rails will determine the extent to which the trestle should open; in the present instance, for a tray 2ft. 8in. long, the trestle should be able to open 2ft. clear, the fillets of the bottom of stand will therefore fit inside. The tray is made out of ½in. wood, the top edges being rounded. The bottoms of the trestle uprights must be cut parallel with the top rail.

A SIMPLE CUPBOARD.

THE cupboard illustrated at Fig. 69 is a useful piece of furniture in the nursery, for children's toys. The measurements given may be carried out if the cupboard is to stand independently in the room, but should it be fitted into a recess, the width will then depend upon the space allowed, and the height and depth can remain as given here.

Fig. 70 shows the cupboard without the doors. The top measures 3ft. 2in. long, 1ft. 6in. wide; the sides 3ft. 6in. long, 1ft. 5in. wide, all of ¾in. wood. The wood is ordinary deal, and on account of the depth the sides, top, and bottom will each have to be made of two boards joined. The top will just take two 9in. boards. After the edges have been planed true, glue the inside edges and clamp them together until quite dry, then finish off with a smoothing-plane. It is advisable when boards require joining, to cut them a trifle longer than will be required. When finished and joined, square the ends, and

FIG. 69.—CUPBOARD.

cut the board the right length. In joining, if the edges are shot perfectly true, glue will hold the boards sufficiently firm to prevent warping, and when planed up after gluing they should have the appearance of one board. This, naturally, requires practice, but can be more surely done by the use of the shooting-board. The bottom should also be glued and clamped similar to the top and sides. If the cupboard is to fit into a recess, the sides need only be planed thoroughly on one side, viz., the inside; the side next the wall and the bottom being just planed roughly with the jack-plane, as only one side is visible. Presuming the top is joined and finished, including the ends squared and cut to the required length, the front and side edges should then be chamfered, rounded, or a moulding run round; or they can be left square, but in this case the cupboard will not have such a finished appearance. Underneath the top screw two pieces of wood 1 in. square, 1 in. from each end, back to front, leaving ½in. at the back and 1½in. at the front; also another piece along the back, ½in. from

the edge; these are for the sides and back to be fixed to. Similar pieces should be screwed on the sides, 3ft. 3¾in. from the top, 1 in. from the front, and flush with the back; these are for the bottom to rest on; and two pieces 1ft. 8in. from the top, for the shelves to rest on.

Having cut the sides the required length, screw them to the slips at the top, and, according to the measurements given, the bottom should be cut 3ft. long, and screwed to the sides, at the same time resting on the slips. The bottom is 1 in. less in width than the side, viz., 16in., and is fixed 1 in. from the front of sides and flush with the back. The division and the shelves are the same width as bottom, and should also be joined and glued. The former is fixed by two slips of wood screwed underneath the top, one on either side, and screwed on to the bottom from underneath; two small slips are then screwed on to the division, similar to the sides on which the shelves rest. The back is best made up out of match-boarding ½in. thick: it is not only cheap, but being tongued and grooved, if

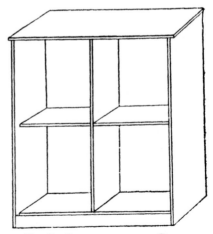

FIG. 70.—CUPBOARD WITHOUT DOORS.

carefully fitted, will not warp. This completes the cupboard, with the exception of the doors and plinth below them.

The doors measure 3ft. 3in. high, 1ft. 6in. wide each, and the frames should be of 1in. wood 2in. wide, the centre stiles 3in. wide. Before mortising the frame together, the inside edges must be rebated ¼in. each way, i.e., ¼in. wide ½in. deep, the centre stiles on both sides. On account of these rebates care must be taken when mortising, as the tenons on the top, bottom, and centre stiles will appear ¼in. longer in the front than they are at the back. The tenons can either come through the uprights, as in the case of ordinary doors, or only three-parts of the way. The latter is prefer-able to amateurs, so long as the tenons fit well, and when glued will hold firmly enough. The panels are best made of ¼in. match-boarding, fixed diagonally, as shown. The writer made a cupboard

with the panels of ordinary deal, i.e., not tongued and grooved, and ran a ¼in. bead on one side of each piece; but this plan is not to be recommended, for unless the wood is well seasoned there will soon be gaps between the boards, whereas the match-boarding pre-vents this. The doors are hung flush with the front edge of the sides with 3in. brass butts, one door is fixed by flush brass bolts, let into the edge at the top and bottom, the other door by a brass button or lock, as preferred. After the doors are hung, fit in the plinth at the bottom, which should be of 1in. wood, so as to come flush with the doors.

The cupboard can be either sized and varnished only, stained and varnished, or painted according to the taste of the amateur.

HANGING OR MEDICINE CABINETS.

FIG. 71 shows a hanging or medicine cabinet. It has two shelves, a cupboard with two doors, and a couple of drawers. The cabinet is 2ft. high over all, and the shelves are 2ft. long. In width from back to front it is 6in. Wood which in the rough is called ½in. thick, though when planed up it will measure little if any over ⅜in., will be thick enough for the end shelves, drawer fronts, and panels. Thickness for other parts will be given when mentioning them later on. The cabinet will look well either in mahogany, American walnut, oak, or ash. None of these are particularly diffi-cult to work, though Honduras ma-hogany, not Spanish, is the easiest. This is the wood that should be chosen if it is desired to ebonise or blacken the cabinet, for with the exception of walnut the others are not so suitable for this finish; they will stain readily enough, but there will be more difficulty in get-ting an even black surface, owing to their coarseness of grain. Whatever the wood,

it should be thoroughly well-seasoned, that it may neither cast (twist) nor split. If, as sometimes happens, a split should show itself when cutting out the wood, do not use the piece which has a flaw in it, but put it on one side till it can be made available for something smaller. For this reason it is advisable in cutting a plank to get the largest parts out first. The two ends, the top shelf, the middle one forming the top of the cabinet, and the two lower ones between which the drawers slide, will be of the same length, viz., 2ft.; but in width the end pieces should be a little more, say ¼in., than the shelves, to allow for the backing of the cupboard and the slip above the top. All of them should be carefully planed on both sides to remove the roughness and inequalities of surface, not to men-tion weather stains acquired while the wood has been lying in the plank season-ing. By the way, if there is any doubt about this being thorough, let the pieces remain in a warm dry room for a few days before fitting them. It is very unlikely that the plane will leave the wood as smooth as it ought to be, so the scraper —a flat, thin bit of steel—should be used subsequently, when a final rubbing with glass-paper will produce an excellent level surface.

The ends, the upper part of which is shown by Fig. 72, must have the edges squared and smoothed, the back and front being parallel. The shaping shown can easily be managed with an ordinary fret-saw, or in the absence of this with chisels. Those who have a fret-saw will have no difficulty in cutting the pattern shown, which, it may be interesting to

FIG. 71.—THE CABINET.

note, is taken from a design by Heppelwhite, and published by him about a century ago. He was one of the triad of furniture designers and makers who did so much to establish cabinet-making as a fine art during the latter half of the eighteenth century.

The pattern will also look well if the spaces are grounded out as for carving, so that the lines of what would otherwise be the fret remain in relief. As an alternative, and one involving less work, the holes may be bored through as indicated in Fig. 71, affording a ready and easy means of simple decoration, the effect of which is by no means to be despised, although it is so unpretending. A ½in. centre-bit

will do all that is needed, though some care must be taken to avoid a ragged edge on whichever may be the under-side when the bit goes through. To prevent the wood tearing, bore from both sides; as soon as the sharp point in the centre of the bit shows through, reverse the wood and place the point exactly in the puncture it has made. With ordinary care the result will be a hole with clean, sharp edges. Next mark on the ends the position of the shelves, the upper one of which is seen to be 3in. from the top. The space between it and the top of the cupboard is 8in. Below this set off 7in., which is the depth of the cupboard, and 2½in. below this again for the bottom. The four shelf-pieces are now accounted for, and to insure their being properly placed the lines marked for them across the ends must be set out with the square. The ends of the shelves must also be made perfectly true and square. All these transverse pieces must be of exactly the same size. It may require time to get them all equal, but the result will compensate. Those who have a shooting-board will not have much trouble, but, even without one, it is very little more than a matter of care and patience to shape the parts properly.

It will be noticed on Fig. 72 that the upper edge of the front of the shelves is bevelled. This may easily be done by planing down sufficiently. The angle of the bevel should be about 45deg., though a little more or less will not signify. The edges of the two lower shelves, viz., those above and below the drawers, will look better if left square. Be careful not to round either them or any other square edges when smoothing down with glass-paper. Nothing looks worse than a rounded arris (the technical name of the angle or line formed by horizontal and perpendicular surfaces), as it is an evidence of slovenly work.

Before fastening the shelves to the ends, these should be rebated behind to let the back be sunk. This rebate should only extend from the top of the cupboard to the shelf below the drawers, unless, indeed, it is preferred to enclose the back of the open space below the top shelf, when the rebate may be carried through from end to end. The small portion of the rebate below the drawers can easily be filled up by glueing in a piece of wood. This will be a simpler plan than cutting a rebate just so far as it is needed.

The recognised tool for cutting a rebate or rabbet, is the rabbet - plane. If the worker does not possess one of

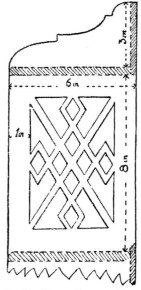

FIG. 72.—UPPER PART OF ENDS.

these, a chisel may be used alone, and though it will be a tedious job it may be managed with care—cutting with the bevel side of the chisel downwards. This will minimise the risk of tearing too much of the wood away. The limits of the rebate must be marked beforehand. What these are must be determined by the thickness of the panel to be let into it, and also by the thickness of the stuff itself.

We already know the thickness of the ends of this cabinet, and reference has been made to that of the back as being about half. Assuming that this is so, the space to be cut will measure say $\frac{3}{16}$in., both in width and thickness

(length is unimportant, except for reasons stated above), so that this space will have to be marked with pencil or otherwise. The best is by means of the marking gauge, which allows the lines to be run with the utmost facility. The possessor of a similar tool—the *cutting* gauge, which by the way will also serve as a marking gauge—will have at his hand a great assistance in forming shallow rebates, as he can cut to a moderate depth by drawing the gauge backwards and forwards with increasing pressure. There is never any difficulty in cutting rabbets $\frac{1}{8}$in. deep by this means, and when they are only a little more, the waste piece can easily be torn out with a chisel, which should be used to trim up the inner angle. The block of the gauge must be adjusted to cut the required width, and be firmly fixed by the screw placed there for the purpose. In the absence of a cutting gauge, which, however, is a very inexpensive tool, and one that comes in for all sorts of woodwork, a small saw may be used.

FIG. 73.—RABBET, WITH PANEL FITTED IN.

In Fig. 73 the rebate is shown in section in the horizontal piece, with the other resting in it.

Fix the shelves with screws—three to each end of each shelf will be enough—and see that the heads are well sunk; *i.e.*, they must not project at all, though they may be without detriment a little below the surface. The reason for this will be seen later on.

We have now got what may be called the carcase of the cabinet together. We may next either make doors or fix the pieces, partly closing the front end to which the doors are hinged. It does not matter which is done first; but perhaps it will be better to fit the two end fixed pieces, as the exact space to which the doors are to be made can then be accurately determined from actual measurement without any chance of mistake. Do not, however, fix them in permanently if they are to be ornamented with the carved panel stuck on as shown in Fig. 71. Presuming that the doors are to be square or as nearly so as convenient, the width of these pieces will be about 4in., and a little calculation will give the exact dimensions. They may be of the same substance as the shelves, and should fit tightly, for if they are too short or have uneven edges, it will be impossible to hide the defect. They may be kept temporarily in position by a needle-point or two driven in from the exposed edge slantingly to top and bottom.

Next make the door framing. For this some stuff about 1¼in. wide by ⅜in. thick will be required, four pieces for the top and bottom rails and four for the uprights or styles, as they are called, of the doors. The length of each piece of the top and bottom of the frame must be half the length of the space to be closed by the doors. The styles likewise must be long enough to fill the space they are to occupy. The more care taken to insure all these pieces being accurate in width and thickness as well as properly squared at the ends, the better; for unless they are it will be impossible to satisfactorily fasten the parts of the frame together. This will be done by

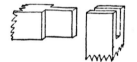

FIG. 74.—MORTISE AND TENON.

the mortise and tenon joint, of which Fig. 74 shows that a tongue or tenon on the top rail fits into a corresponding hollow or mortise in the style. They should fit tightly to each other, but not too much so, or the tenon is apt to split the mortised rail open.

The following details of mortising may be useful. There is a special gauge for this construction, but the ordinary kind will do very well, so let us suppose it is to be used. Set it to the width of the frame, then mark this distance from the end on each face of the upper rail as shown by the dotted lines A on Fig. 75. Do the same on the style, but instead of marking all round, it will only be necessary to do so on the edges,

not on the back or front (B, Fig. 76). We have now got the length of tenon and depth of mortise, which if cut to the lines must be equal, but the thickness

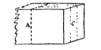

FIG. 75.—MARKING FOR TENON.

of tenon and the corresponding width of mortise have still to be ascertained. The gauge must again be set, this time to about one-third of the thickness of the wood. Then mark with it, from

FIG. 76.—MARKING FOR MORTISE.

each side alternately, the lines C, C, on both the rail and the style, when it only remains to cut away the outer blocks from the former and the centre one from the latter. The saw alone is required to form the tenon, but it can only do part of the mortise, and the waste piece must be chopped out with a small chisel. With care, an ordinary chisel will do very well, but there is a specially thick kind called a mortise chisel, the advantage of which is in its greater strength. Any small saw will do, and as guides for its course, the lines marked by the gauge must be closely adhered to, though due allowance must be made for the kerf or passage of the saw by keeping to the outside of the lines marking the tenon and the inside of those for the mortise. The novice must not be discouraged if after all he find the joint not quite so neat as he would like, and he may be reminded that it is always easier to remove a shaving if the fit be too tight than to fill up a loosely-made joint.

When the door frames are made, grooves should be made on their inner edges for the panels to fit into before they are glued together. If a plough is not available, the cutting gauge will

E

answer here, the superfluous wood being removed with a chisel. The panel may be about the same thickness as the shelves, and of course it must fit the groove closely. The section of the frame with groove and panel fitted in is shown in Fig. 77. It will add to the appearance of the door if the upper and lower rails are bevelled as indicated instead of being left square. The panels may also be left plain if preferred, but some simple device carved on them (as shown in Fig. 71) will be a great improvement. The carving should be done before the panels are fixed in the frames, as it will be more easily managed. It will be better not

FIG. 77.—SECTION OF FRAME, RAIL, AND PANEL.

to use any glue to fasten the panels in, as the grooves alone will hold them. The frames must, however, be glued at the joints, which should be tightly cramped up that the parts may fit closely together without leaving any ugly space at the joints.

Before beginning them, however, the upright division should be fixed. It may be the same thickness as the shelves, and must fit in tightly; probably some odd piece will come in handy here, for the width is not important, but it should not be less than 1in. The grain of the wood, whatever distance the piece extends back, will of course be perpendicular, as it would be quite opposed to practice for it to show "end grain" in front. This piece, though it would keep the drawer fronts apart properly, affords no support or next to none for the drawers themselves, especially if it is only narrow, say anything under 3in.; so a rail separating them their entire length should be placed on the bottom behind this central piece. The rail may be of pine or any other wood, but it must be carefully prepared and fixed, for it acts as guide for the drawers. Its thickness must be the same throughout, and equal to that of the upright piece in front. Its width is not so important, but there is no necessity for it being

more than ½in. or so. This slip may be glued on, taking care to remove any of the glue which exudes, before it hardens. The front piece may be easily fastened

FIG. 78.—DIVISION BETWEEN DRAWERS.

in by a couple of small screws or brads driven in from above and below. Fig. 78. shows this part, A being the bottom of the cupboard, or shelf above the drawers, and B the shelf below them, in section. Between these are C, the upright piece in front, and D, the strip running to the back.

When ·these are fixed, prepare the drawer fronts. These should fit tightly into their respective places, and may be the same thickness as the shelves, though a little more or less will not signify. The pieces for the drawer sides should also be prepared. They must be made to fit in their respective places as closely as possible, especially in regard to width, but it will be better to cut them, say, ½in. less in length than the full distance to the back of the cabinet. The thickness of these pieces need not be more than ¼in., and it is not at all unusual to make them as well as the drawer backs and bottoms of a different wood from the front. Honduras cedar, the kind cigar boxes are made of, is often used for the purpose, though a good clean piece of pine is equally serviceable. The drawer back may be of the same size as the front, but if originally got out so it will have to be lessened in width afterwards, so it will be just as well to let it be ¼in. to ⅜in. narrower.

We now come to the somewhat difficult operation of dovetailing, to which the beginner can hardly devote too much care if he is desirous of becoming a competent joiner, for it is a joint which, either in its simple form, as in this drawer, or in its more complex varieties, is constantly used, and nothing betrays the unskilled workman more clearly than a badly fitted dovetail. The dovetail, as about to be described, is the ordinary plain dovetail which is used whenever it is not objectionable for end grain to show. Where the end grain would be unsightly, the "lap" dovetail is generally preferred, but in the present case the ordinary dovetail will suffice.

Presuming that all the pieces, except the bottom, have been prepared, properly squared up at the ends, &c., proceed as follows: Set the gauge to the thickness of the front piece, and mark the front ends of the side pieces from top to bottom on both sides. Then set the gauge to the thickness of the back piece, and mark the other ends of the sides in the same way. Now, if the drawer sides and back are of the same thickness, the gauge may remain as it is for marking the back and front; otherwise the gauge must be reset, but this time to the thickness of the sides, which must be marked off on the ends of the front and back. In all cases both sides of the wood should be marked, and the gauge should be set, if anything, rather

FIG. 79.—DRAWER FRONT DOVETAIL PINS.

less than beyond the thickness of the respective parts by which it is regulated. Any difference should, however, especially in the hands of the novice, be as little as possible; still, if there is a deviation, it should be as indicated, for less harm will then result; indeed, the skilled workman often with advantage marks within the required space, for the reason that by great accuracy in cutting, he is able to produce a closer fitting drawer. Now on the drawer front cut the pins, the number of which is not important; but in a small piece of work like the present three will do very well, as shown in Fig. 79. The spaces between them, represented by the shaded portions, are to be cut away as far back as the lines marked by the gauge. Naturally the pins must be marked on the wood before cutting, especially on the ends. From the ends lines may be marked

at right angles, by the aid of the square, to meet the gauge line; but this is hardly necessary, except for the beginner, who may be guided by them when sawing. From this it will be gathered that the saw, which should be a small one, must be used to cut down the gauge line. The saw should work outside the pins, in order that these may be left full size, and when cutting the corresponding parts, viz., the sockets for them, in the drawer sides, on the inside of the line. By this means the saw cut is compensated for, and the two parts will fit each other tightly.

Having sawn down to the gauge-line, chop out the pieces with a chisel, but in order that a clean edge may result on both sides do not cut from one only. For this reason chiefly the wood is gauged on both sides, and the chisel should cut slightly inwards from each surface. We have now got the pins which are to fit between the dovetails on the side pieces. These must be marked in order to insure their being in proper position, and there are several ways of doing so. None, however, will be easier than one of the methods very generally adopted by practical workers. It is this: Place the end upright on the side piece, which we may suppose to be lying flat on the bench. See that the two pieces are in the proper relative positions they will occupy when the drawer is finished by noticing that the outside of the front is flush with the end of the front, or, what is the same thing, that the back of the pins are on the gauge-line of the side piece. It will be also necessary to note that the edges on top and bottom of the two pieces are even. When adjusted, hold them so and mark the position of the pins on the drawer side either by scribing with a sharp point or with a

FIG. 80.—DOVETAILS ON DRAWER SIDES.

fine lead pencil. This will give exactly the position and shape of the spaces into which the pins must fit, or, in other words, will give the shape of the dovetails which will fit into the spaces cut for them in the front (Fig. 80).

The pieces should be chopped out with a chisel as described for the front. It will now be found that the two pieces when fitted together form a firm joint

FIG. 81.—DOVETAIL JOINT.

(Fig. 81), which can be disconnected by a lateral but not by a forward pull. The same proceedings are taken with all the other parts.

Before proceeding to fit the back of the drawer it will be well to prepare the grooves for the drawer bottom to slide into. Here again the gauge will come in handy, as with its aid the position of the groove on the front and sides may be exactly determined. It should be, say, ¼in. above the bottom edge of the drawer, but its exact position is not material. If too near the edges there will not be sufficient strength, and if, on the contrary, it is too high up, the containing space of the drawer will be needlessly curtailed. The width of the groove must be determined by the thickness of the drawer bottom, which, unless the sides are very slight, as in the present, is usually the thinnest part of a drawer. It is, however, better to cut the grooves rather narrower than the thickness of the bottom, as this can easily be reduced at its edges to a proper fit by bevelling them, preferably on the underside. The depth of the groove should only be slight and not be such as to weaken the drawer sides too much. When the groove is made we get at the position of the bottom edge of the drawer back, which should be just level with the upper edge of the groove in the sides, so that when the bottom is put in there will be no opening between them.

If cutting a groove in such thin stuff as the present drawer sides is considered too risky by the "'prentice hand"

—and this part of the work requires rather nice management—here is an alternative, which if not quite so neat, is if anything stronger, and adopted by some of the best cabinet makers. Instead of cutting a groove in the side, form a space for the drawer bottom to slide in by glueing a strip along the bottom of the sides and another higher up, the space between the two being equal to the thickness of the drawer bottom, which, as in the case of grooves being cut in, may be bevelled off to fit (Fig. 82). The upper slip, which it will be noticed is within the drawer, looks better if rounded as shown. It should entirely fill the space between the back and the front of the drawer, but the lower slip should come right to the end of the side and be cut off level with it, as it thus forms a cleaner-looking job than if it were shorter. The drawer bottom should be cut so that the grain runs across from side to side, not from

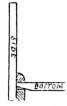

FIG. 82.—DRAWER SIDE WITH BOTTOM FITTED WITH SLIPS.

back to front, to allow for any probable shrinkage. For this reason it should not be glued in, but be merely pushed in the groove, and it will also be well to allow it to project a little—say ¼in.—behind. It may be secured by a small screw driven through into the back. Unless made of the driest wood obtainable, and even then there may be some risk, the drawer bottom is very likely to contract. Not being glued in it will not split, but it will recede from the front and leave a gap there. To close this, the bottom must be pushed forward from behind. It is easy to withdraw a screw to allow of this being done, but to get a small brad out neatly would be another matter altogether. Hence the preference for screws as a fastening for drawer bottoms, and the reason why there should be a projection behind in all, or nearly all, new drawers.

With the exception of the bottom, all the parts of the drawers are fastened together with glue.

From Fig. 71 it will be seen that the drawers have a panelled appearance. This is arrived at easily by shaping a thin piece, say ⅛in. thick, or even a piece of veneer, as shown in Fig. 83, and in a slightly more ornamental form in

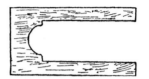

FIG. 83.—PIECE FOR DRAWER FRONT.

Fig. 84. Those who possess a fret-saw will have no difficulty in doing this, while others will not experience much in cutting it with other tools. The pieces are fastened on with glue, and in cutting them it will be as well to let them be a little full—*i.e.*, larger than the fronts, to which they may be trimmed down after the glue has

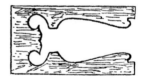

FIG. 84.—ALTERNATIVE FOR DRAWER FRONT.

hardened. These pieces laid on serve not only as a simple means of relieving the monotony of a perfectly plain drawer, but also hide the ends of the sides.

The pieces between the doors and the ends have a panel planted on in front and fastened with glue, and a couple of small screws from behind. The panels are ⅛in. thick, but wood of the same substance as the ends will do very well. The edges are bevelled to half their thickness, and the carving is of the simplest character, being nothing but a series of round buttons left by the grounding out, which is done to the

depth of about ⅛in. Here, again, fretwork may be substituted if preferred. If it is intended to carve these parts, the screws referred to should not be put in

FIG. 85.—BLOCK IN POSITION.

till the carving has been done. Glue alone will hold firmly enough till the carving is done. When the carving is done, the pieces to which they are attached may be fastened to the cabinet. They should be set back about ⅛in. from the front of the ends, and fixed by blocks glued into the angles at top, bottom, and side. The blocks, for the sake of neatness, should be triangular in section (Fig. 85). In order that no excess of glue may be left between the contiguous parts, put the blocks in with a sliding movement rather than by direct pressure. In addition to the block fixings there will be no objection to the edge being glued to the end; but if this is done the blocks need only be placed at top and bottom. Much, however, will depend on the accuracy of the work, as if all the parts fit closely, they will not require the same support that they otherwise would from glue. The doors are attached by a couple of small butt hinges to each. Mark off the places on the edges of the doors, and then cut away sufficient to allow the hinges to be sunk in them level with the wood, in order that the doors may fit closely. The hinges are fastened with screws, the heads of which must not project. The

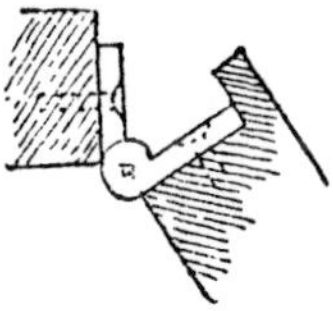

FIG. 86.—HINGE ATTACHED TO DOOR.

knuckles of the hinges must project a little to the front, and as a general rule to be followed, it may be stated that the pin connecting the flanges of the hinge should be just clear of the front of the door (Fig. 86). If the hinges project a little more than is absolutely necessary in front, the only objection, apart from the unworkmanlike effect to a critical observer, will be that the doors will throw open further than they otherwise would.

The best way to attach the doors is first to fasten the hinges to them, and then, holding the doors in position with the hinges open, to bore one screw-hole through each hinge plate into the adjoining wood. Then fix the door, and if it hangs all right put in the remainder of the screws. If it does not, take out the screws and try again with another hole. This is a much better way than boring all the holes before making a trial of the hang of the doors, for in the event of a misfit, the fact of one hole for each hinge being a trifle irregular is not of particular importance. If the door really fits accurately, it is a good plan to put something, such as a piece of card or glass-paper, under it when holding it to bore the holes in the wood to which it is fastened, in order to obtain just sufficient space under the door to prevent its dragging on the shelf below. For the same reason it is necessary that the hinges should be firmly screwed home, otherwise the door is almost certain to drop.

Before the left-hand door is fastened a small bolt must be attached to it. Two kinds are suitable, the flush bolt, which is sunk on the edge, or the blind bolt, which is simply screwed on behind. The flush bolt may be fixed either at the top or bottom, but the other should be at the bottom, and as near the edge of the door as convenient. A hole to receive the bolt must be made in the bottom of the cupboard. Its exact place may easily be ascertained after the door is fixed by covering the under end of the bolt with a little gas black or anything that is convenient, closing the door and forcing the bolt down on to the wood, when an imprint will remain showing where the hole is to be bored. This can be left plain or protected by a small piece of brass, which can be obtained by asking for a bolt-plate to fit it when purchasing the bolt, or it can be made by simply boring a hole through a thin piece of brass, which of course must be

sunk level with the wood to which it is screwed.

On the right-hand door a lock should be placed, or a "cupboard turn" will suffice. This is a kind of catch of which only the handle is visible on the outside; but perhaps a better fastening is the "bullet catch," as the mere opening and closing of the door actuates the bolt. This is spherical at the outer end, and is kept projecting by a spring, which, however, is pushed back whenever the bolt comes in contact with the bolt-plate, whether in opening or closing the door. A bolt of this description may be placed either in the upper or lower rail of the door, or in the ordinary position of a lock. Being let into the frame, it is not visible from the outside. In connection with it, though not necessarily on the same part of the door frame, a small handle of some sort will be wanted. A small brass knob, with a fancy head and screw shank, should, for the sake of appearance, be chosen. Melhuish and Sons show some very suitable designs.

The drawers also will require handles. They may be similar to that on the door, but a comparatively plain one will suit the present cabinet better than any of the highly elaborate devices. If a lock be used, let it be a "cupboard lock," with bolt shooting to the left." When ordering, the width of the plate (which should not be wider than the frame of the door) should be mentioned. The keyhole should be covered by a small "escutcheon plate," and it should not be larger than is required to pass the key, but the hole in the wood should be at least as large as that in the plate.

On the right-hand door glue a small slip of wood, which will look all the better if it is bevelled on the front, or beaded, so that when the doors are closed half of it overhangs the one on the left hand. This slip gives a better finish to the work, and is especially useful in concealing any gap there may be through shrinkage or defective workmanship.

Behind the left-hand door, a small stop, 1in. by 1in. by $\frac{1}{4}$in. thick, should be glued to the top or bottom, with, perhaps, a brad or two in case of the glue giving way. Similar pieces may also be useful to prevent the drawers being pushed too far in, if they do not happen to go right to the back, as they very seldom do. The stops may be placed quite behind the drawers, fixed on the shelf or ledge on which the drawer runs, just sufficiently far back to catch the drawer front. When this is the case, of course the stop must be thin enough to allow the drawer bottom to slide over it, and it must be out of the way of the drawer sides.

The back of the cupboard has still to be fixed. It is simply fastened with a few brads into the rebate prepared for it. Unless it shows above the cupboard, a piece of thin pine will do as well as anything else, and if it be thoroughly dry, so as to prevent any chance of shrinkage, it may be in one piece, the grain of which should be—and this is a safe general rule—in the direction of the greater measurement. In this back the grain will be horizontal. The slip above the top shelf is fixed in a similar way, but it should not be pine, unless the whole of the cabinet is of this wood.

The two shaped pieces are the only parts that have not been mentioned. They must fit closely in the spaces where they go, the straight edges and ends being nicely trued up. Glue alone will hold them; but it will be better to put a block or two behind them similar to those behind the pieces at the ends of the cupboard. The corners of the upper rail may be lightened by boring holes through them, as shown in Fig. 71, or they may be cut to match the fretted ends.

In small cabinets of this description, where the shelves are merely screwed to the ends, it is usual to hide the screw-heads by gluing over them small buttons either turned or shaped as in Fig. 87. These latter can easily be formed by

FIG. 87.—BUTTON TO COVER SCREW-HEAD.

bevelling the edges of thin ($\frac{1}{8}$in.) wood with a chisel; but whatever the shape, they should be large enough to extend well over the screw-head and allow the glue to act on the wooden surface of the end. Nothing now remains to be done but to polish the cabinet.

Fig. 88 is a front view of another style of cabinet with one door removed; it

is divided into two by a shelf, and at the bottom are two drawers for boxes, &c. The cabinet may be made of oak, ash, or any hard wood polished, or of pine varnished. The lower portion is 11in. high, and the upper portion 9in. high; the depth inside is 6½in., the width 23in.,

FIG. 88.—FRONT VIEW OF HANGING CABINET.

and the depth of the drawers 2in. The two ends are 8½in. wide and the top and bottom 8in. wide and 1in. thick; the full length of the ends (Fig. 89) is 3ft. The top and bottom of each end are cut as shown, and a bead is worked on each edge, stopped about 1in. from top and bottom; grooves are cut in them for the top, bottom, and inside shelves. The top and bottom have a bead worked on each edge carried right through to the end; the ends have a rebate cut on the inside of the back edge ½in. each way, for the back to fit into. The back is formed of two or more ½in. boards jointed together, the ends being cut as shown.

The wood must be thoroughly dry and well seasoned, and as soon as the various parts are fitted together they should be brought into the house and set in a dry place for a few days; then glue up the top, bottom, and sides, and fix the back by screws into the top and bottom and

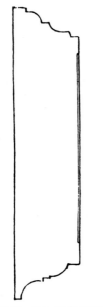

FIG. 89.—END OF CABINET.

shelves. Then make the two doois. The rails are 1in. thick, and the panels ½in. thick. A bead and two hollows aie worked in the centre of each rail. These and the stopped beads can be worked by a "router." The rails of the doois are mortised and glued together, and the panels are thinned at the edges and fixed into grooves in the rails.

Fit the doors into the front, and hang each by two brass hinges (the hinges must be let in flush); fix two flush bolts in the edge of one door, and a lock with an ornamental brass cabinet handle in the other door. A beading about ½in. wide must be fixed on the edge of the door with the lock in it, to hide the joint. Fix to the top of the cabinet, by screws, two brass plates with holes drilled in them, for the purpose of hanging it to the wall.

Two drawers are fixed in the bottom of the cabinet inside; the fronts are ¾in.

thick, the sides and back ⅜in.; the joints are dovetailed at the coiners, and the bottom is thinned at the edges, fixed in a groove in the sides and front, and fixed to back (which only comes to the top of the groove) by two or three brads. Make the drawers an easy but not slack fit, and let into the front a brass flush handle; fix cross divisions in the drawers, with the top edge bevelled, and divide between these into compartments

FIG. 90.—SECTION OF DRAWER.

for boxes, pots, &c,, as shown in Fig. 90. Labels can be stuck on the bevelled parts, with the names of the contents of each division.

Fix about 2in. above the bottom shelf, rods ½in. square, between which the bottles are set, the largest at the back as shown on Fig. 91. If each bottle is labelled, the required one can be found at once. On the top shelf fix a thin board, 1½in. above the shelf, with rows of holes to suit 2oz., 1½oz. and 1oz. bottles, setting the longest at the back, so that all the labels can be seen.

Bottles containing poisons should not be put among the others, but a space

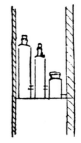

FIG. 91.—SECTION OF CABINET.

should be partitioned off, and a cover fixed in front with two biass buttons and a large label on with "poisons" printed on it.

An oiiginal design for a medicine cupboard is shown in Figs. 92 and 93. Referi ing to Fig. 93, which shows the interior of the cupboard, a general idea is given to the amateur before proceeding to make it. The division

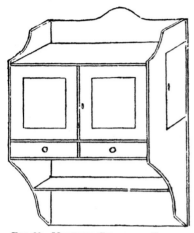

FIG. 92.—MEDICINE CUPBOARD, CLOSED.

on the left hand is for large bottles (in arranging them always put the tall ones at the back): this extends the whole depth of the cupboard. The

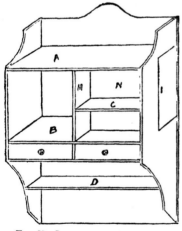

FIG. 93.—INTERIOR OF CUPBOARD.

one on the right hand is half the depth, and further has a shelf in the middle for small bottles: the back half

forms a private cupboard accessible by a small door in the side, as shown, for the storage of poisonous medicines, which can be kept under lock and key. It is unfortunately but too well known how fatal accidents have arisen through persons mistaking the wrong bottle in taking medicines. The plan of keeping such medicines in a separate compartment will obviate these fatal errors; besides, they will be out of the reach of enquiring children. The two drawers beneath the cupboards are suitable for storing plaisters, pills, and such-like things, and these can be simply secured by inserting a small pin of brass or iron wire through the bottom of the cupboard

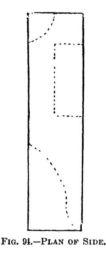

FIG. 94.—PLAN OF SIDE.

into the front of drawers. The shelves top and bottom can be utilised for keeping books on or for ornamental purposes.

The material used in making the cupboard need only be of the cheapest, the idea being that when finished it should be enamelled a colour to suit the taste of the amateur, or to harmonise with other things in the room where it is intended to be fixed.

The extreme height of sides is 2ft. 3in , depth 7in , $\frac{1}{2}$in. thick, the straight edge in front is 12in. starting 4in. from the top, which is hollowed out as shown. Fig. 94 will give a better idea how to prepare the sides, and only on one side is the door to be cut out. If care be taken in cutting out the door for the side, the piece of wood cut out can be utilised for the door. The length of cupboard between the sides is 18in., depth 7in. (same as sides). The height of the cupboard inside is 8½in , of the drawers 2in. By way of finish the edges of the sides can be much improved by cutting a small bead or flute on them: this can be easily done by the aid of a bead-router which is made in the form of a spokeshave.

Having cut out the sides, fix the top and bottom of the cupboard (A and B) to them with thin screws, countersinking the heads well in ; then prepare the centre partition (M), which should be 6½in. from back to front, standing back ½in. from the front to allow for the doors. Before fixing this it will be found advisable to fix the middle partition (N), and the shelf (C) to same. The cupboard being 7in. deep, the partition (N) is to be fixed half way, and should be screwed to partition (M), and to the right side ; the shelf (C) to be screwed to partitions (M) and (N) and to the side. Then, having fixed the two partitions and the shelf together, proceed to put them into position by screwing the partition (M) to the top and bottom of the cupboard.

The bottom of the drawers to be then screwed to the sides 2in. below the bottom of the cupboard, the division between the drawers to come out flush with the front, and if well screwed from the bottom will be sufficiently firm without further fixing, unless the amateur prefers to glue the top

edge to the bottom shelf of the cupboard.

The back, which should be of ½in. wood, can be made in one or two pieces, and is screwed on from the back. The under-shelf (D) is fixed 5in. from the bottom of the drawers and should be 4in. wide. The small door at the side is hinged on to the back, and a thin slip of wood is fixed inside the side to act as a stop, in order that the door may shut flush with the side. This door should be so well fitted as to be barely noticeable when shut. This will complete the cabinet so far as is shown in Fig. 93, the only remaining parts being the doors and drawers.

The doors should fit flush with the front, the middle partition (M) acting as a stop for them. They should be 8½in. high, 9in. wide, and ½in. thick, a small slip of wood being fixed on one, so that when both are closed the one cannot be opened without the other. A small frame, 1½in. wide and ¼in. thick, is glued and beaded on the face of each, not only for appearance, but to prevent warping, which otherwise would assuredly happen, the doors being made out of one piece. The inside edges of these frames can be left plain or beaded—the latter is recommended.

The drawers, allowing for the middle division, are 8½in. long, and 7in. back to front. The front should be ⅜in. thick, the sides and bottoms ¼in. or ⅜in. One knob to each drawer, either in brass or wood, and a small lock and key for the doors, will complete this useful cupboard. Size the wood before enamelling, which will prevent the enamel drying in.

SIMPLE SCREENS.

IN our grandfathers' time screens were common enough, but those worthy ancestors were not æsthetic : they used their screens simply as screens to ward off draughts, therefore they covered them with red moreen or baize, nor were they content with anything less than 8ft. or even 10ft. high, and from four- to six-fold. Nowadays the screen has returned to favour, and is to be seen in almost every drawing-room; it is made in almost every possible shape and composed of all sorts of materials.

The present subject is of the very simplest construction, and the materials employed are inexpensive. The amount

FIG. 95.—DESIGN FOR LEAF OF SCREEN.

of mechanical skill required is moderate, and if the directions be followed, the result will be a most presentable and

perhaps a handsome screen. Should the amateur not be an artist, the screen may be made beautiful by adapting some of the numerous art fabrics, etc., which form so distinctive a feature of the present day.

The directions are confined to the making of one leaf or fold. The screen may, however, be two-, three-, or four-fold, according to the fancy and requirements of the maker. The height will also be a matter for individual decision.

The wood employed for the frame-work is red deal; the moresque arch is of oak as used for fretwork; and the panels are Willesden paper, either hand-painted or covered with Liberty cretonne or Lincrusta Walton.

The proportions of the screen, a leaf of which is represented at Fig. 95, are as follows: Extreme height, 4ft. 9in.; panels, 2ft. square; while the frame-work is of stuff 1in. square throughout. For the side uprights prepare two pieces of red deal, 4ft. 6in. long and 1in. square, finished work. For the three crossbars and short upright for the centre of the lowest portion prepare four pieces, 2ft. 2in. long and 1in. square. The wood should be dry and clean-grained, being free from knots, &c., and all should be carefully planed and accurately squared. The top and centre crosspieces are to be mortised into the uprights, thus leaving the latter 2ft. apart, while the lowest crosspiece may be cut exactly 2ft. long, and be secured at each end through the uprights by two slender wire nails about 2½in. long; if the latter are carefully driven, and the above precautions as to measurement are observed, this will make a joint sufficiently strong for our purpose when further strengthened by the brass corner-pieces (Fig. 96), to be particularly described presently. The short centre upright may be fixed in the same way, or, should the worker prefer it, it may also be mortised to the second and third crosspieces.

The mortises and tenons should be carefully cut, and firmly glued and

wedged. The fact that the ends show
upon the sides of the frame need not
cause any anxiety, as this is provided
against by the small brass plates (Fig
97), which not only serve to hide these
discrepancies but also add considerably
to the appearance of the screen. These
plates are not, however, to be taken in
hand at this stage of the work.

When the framework is thus complete, it should have a coat of Aspinall's

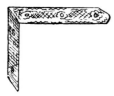

FIG. 96.—BRASS CORNER-CLASP.

white enamel laid on pretty thickly,
and be placed aside to dry. When
thoroughly dry, it should be rubbed
down with very fine glass-paper wrapped
round a flat piece of wood, care being
taken not to interfere with the sharp-
ness of the corners of the work.

The next items to engage our attention are the panels, one 2ft. square, and
two 2ft. long by about 11½in. wide, or
all three panels exactly fitting the open-

FIG. 97.—BRASS PLATE.

ings, whatever the size of the screen
may be. For our present purpose there
is no better material than Willesden
Paper, either B180 or No. 320 "Excelsior." The former, being the colour
of ordinary brown paper, will need to
be painted in some suitable colour, but

the latter, being itself a pretty shade
of green, might have some decorative
design executed upon it, without the
necessity of painting a ground. These
remarks are merely suggestive, but this
paper cannot be too strongly recommended as a material most suitable for
decorative purposes, being strong and
light, while it is unaffected by moisture, and the surface is just smooth
enough and yet rough enough for hand-
painting.

As suggested in the sketch Fig. 95,
a simple design from some of the many
Japanese books may be utilised for our
purpose, and if the drawing of these
subjects is accurately copied, no great
skill in colouring is necessary to make
very effective panels.

The moresque arch mount is of oak
about ⅜in. thick; each arch may consist of two pieces, joined or touching
at the top. Paint these in two coats of
white enamel, rubbing down each coat
until quite smooth. A very pretty and
novel effect may be produced upon oak
treated in this way by going over all the
lines of the grain of the wood—when
the paint is perfectly dry—with a steel
pen dipped in rather thin gold size.
When the size is at the right stage of
"tackiness," apply gold leaf, and the
result will be, upon brushing away the
superfluous gold, that the grain of the
oak will appear in gold, which, in contrast with the white, is highly effective.

The lower panels may be treated in
the same way as the upper one, but a
better contrast will be effected by covering them with Lincrusta Walton, painting it also with white enamel, and
slightly gilding the more prominent portions of the design.

Both sides of the screen may, of
course, be made the same, or rather
treated in the same way; but if it is
required to show only one side, the
back of the panels may be covered with
some of the cheaper Japanese leather
papers, when, although it will not
bear comparison with the front, it will
be quite respectable.

The method of fixing the panels,
arch, &c., is shown at Fig. 98, where A
is a section of the frame, B represents
the Willesden panels, C C the moresque
mounts, and D D a small bead or moulding, either gilded or painted. A mould-

ing which is well suited for this purpose may be had from H. Zilles, Wilson-street, Finsbury. No. 8 in his catalogue is just the thing required. The moulding is fixed with very thin wire nails, or with common pins, in either case cutting off close when driven sufficiently far to secure the bead. In Fig. 98 provision is made for both sides to be the same, but in case only one side is finished with the arch, the bead may be placed in the angle next the panelling.

The open-work ornament upon the top of the screen can be made of any hard wood, ¼in. thick; it also is fixed by means of the small moulding on both sides.

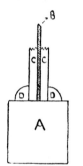

FIG. 98.—METHOD OF FIXING PANELS, &c.

The turned ornaments upon the top of the sides hardly need any special directions. They should be 1in. in diameter at the thickest part, and be painted the same as the frame, to which they may be attached by means of a small dowel and glue.

To make the corner-clasps Fig. 96, cut a piece of sheet brass, No. 20 B. W. C. T., 6in. long and 1in. wide. Bend it at right angles in the centre, and drill holes for small screws or brass nails, as suggested in the woodcut. The part which shows upon the side of the frame may be ornamented in the way pointed out in the article on "A Corner Bracket Cabinet." The same remarks apply to the plates Fig. 97, one of which will be required to cover each mortise end.

Having made the desired number of folding leaves, they may be joined by brass butt hinges placed about 6in. from the top and bottom of the sides.

The following screen is a triple-fold one for scraps. For very large rooms

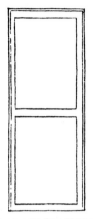

FIG. 99.—FRAME OF ONE FOLD OF SCREEN.

four or five folds are used, but the three-fold is most commonly seen. Fig. 99 shows the frame of one fold; the width of which is 1ft. 8in., and the height 6ft., which is a fair average size.

Procure six pieces of pine 6ft. long, 1½in. by 1in., and nine pieces 1ft. 8in. long, 1½in. by 1in.; make these into three frames similar to the one shown at Fig. 99, plane the pieces up to one size, and mortise the joints, gluing and pinning them. Fix to the top and edges of the frames a mahogany bead 1½in. by ¼in., with a bead on each edge

FIG. 100.—FOLDS OF SCREEN.

of the mahogany, except for the inner folds, where the edge of the mahogany must be flush with the pine, the canvas covering being carried across the joint as shown in Fig. 100. Now fix the three frames together by three brass hinges between each, let in flush, to fold as shown in Fig. 100. Cover each frame on both sides with canvas stretched as tightly as possible, and nailed close to the beading by small, flat-headed tacks, carrying the canvas of the two inner folds across the joint, as shown by the black lines on Fig. 100; then give the whole of the canvas two coats of size or paste, which will cause it to shrink, making it as tight as a drum. When the size is dry, cover both sides of each fold, close up to the beading, with plain paper of any colour to taste, for a background for the pictures—blue, red, black, or any colour may be got from the paperhangers; it must be pasted on level, and without any air-bubbles being left under it. When this is dry, the pictures may be affixed; they should be cut out and pasted in groups, or in any way to suit the fancy of the maker.

Coloured pictures are more effective for screen-covering than plain ones, and if large pictures are used they should be in the centre of each panel, with smaller ones round then. A very good panoramic effect may be got by having a view of a town on a river, at one corner, carrying the river on once or twice across the screen; on the banks figures, houses, castles, &c., may be put, with boats on the river. A procession of walking and riding figures is also very effective; but every maker of a screen has different ideas of what looks best. After the pictures are all on, a piece of narrow bordering-paper should be put round each fold, and the pictures sized and afterwards varnished with map varnish.

Sometimes the two middle joints of screens are made as shown in Fig. 101,

FIG. 101.—RULE JOINTS.

but this joint entails a good deal of labour, and special plans are required for the moulds, and it is not much better than carrying the canvas over the joint.

The mahogany beads should be either varnished or French polished.

SIMPLE WINDOW-BOXES FOR PLANTS.

WHAT will be described here are rough-and-ready boxes perhaps —but none the less presentable so far as appearance is concerned. The strength will naturally depend on the way they are made; that is to say, on the manner in which the pieces are fastened together. An expert joiner will almost naturally say that the corners must be dovetailed, but nails can be used instead of dove-tails, and a properly nailed box is at least as rigid and strong as if it were put together with badly fitting dovetails.

Naturally the length and width of the box will depend on the size of the window-sill on which it is to rest, the depth being regulated by the size of plants, but 6in. to 8in. is enough for an ordinary window. Any common wood will do very well, and stuff from old packing cases may be suggested as useful. The thickness should not be less than ¾in., unless for very small boxes. For long boxes the wood will be better if about 1in. thick. Five pieces of wood will be required for the sides, ends, and bottom, for each box must be the same width, each pair being the same length, and of course sawn off square at the ends. If the wood has been got from old packing-cases it may hardly require any planing to make it smooth, except, perhaps, along the top edges. Those at the bottom may be left, as it will not signify if they are a bit rough. Whether from new or old material, it is not necessary to plane down the inside of the pieces, as it is quite sufficient if the outside is cleaned up.

Plain window boxes to cover with cork, &c. it is not necessary to plane up either the inside or outside. The back and front had better be nailed on to the ends—*i.e.*, the ends will be between the other pieces and not on them. It is in any case the better construction of the two, but if the box is to be covered with cork or other material it is hardly so important if the lengths of wood do not cut up so advantageously. It is, however, seldom that window-boxes must fit so exactly that an inch or so smaller than the available space is of any importance, so that, unless there is some very good reason for the alternative plan being adopted, the ends should be placed inside the back and front. To fasten them, French nails will be found the most convenient, as they are easily driven in, their pointed ends rendering it unnecessary to bore holes for them before insertion. Any nails, however, will do, but to use screws would be a waste of material and time, as they do not hold to the best advantage on end grain. Supposing the wood to be of the substance named, 2in. nails will do, though it will be better for them to be

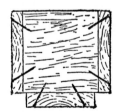

FIG. 102.—METHOD OF DRIVING NAILS.

longer rather than shorter. Now, if all of them were driven in straight, a very slight pull on the front and back would separate these parts from the ends; but let them be driven in "on the skew" or slantingly, as shown in the diagram (Fig.

102, and then try to pull the box apart. It can hardly be done. The wood will give way almost before the nails can be withdrawn from the ends. Practically the box could scarcely be stronger, and, if it were, the strength would be superfluous. Three or four nails at each corner will be ample, but one should be moderately near the top and another near the bottom, unless the bottom board is nailed to the back and front pieces as well as to the ends. If the box is not to be covered, the heads of the nails should be punched in, the holes being filled up with putty. The bottom board may be the same length as the front, so that it goes under and not within the ends, or within the ends if preferred; one plan is as good as the other. Whichever way the bottom is fitted, it is simply fastened with a few nails, and if there is any prospect of moving it at any time when filled with earth, be careful that the nails hold well.

Some provision must be made for excess of water to drain off. This may be accomplished by boring holes in the bottom with a centre-bit (not a gimlet) or red-hot iron, by leaving a space of say ½in. between the bottom and the back and front, or by making the bottoms of two or three pieces, which may have a small space between each. By nailing the pieces across from back to front instead of from end to end, short bits scarcely useable for any other purpose may be made to serve. If they are not of regular width, it will be better to nail them within the box rather than below it, for the sake of appearances. The box now only requires decorating to render it fit for use.

Cork has a pretty rustic effect. That required is "virgin," or cork in the rough bark, and it is to be obtained at a very low price from cork-cutters, many seedsmen and nurserymen also keeping it. No skill is required to fix it; but judicious attention to arrangement of pieces is desirable, and anything like stiffness must be avoided. The best way is to fasten the larger pieces on first, and then fill up the spaces which there will inevitably be with smaller ones. French nails are the best to fix the cork down with, and, as a rule, it may be said that the appearance is better when the pieces are arranged with the wrinkles or cracks perpendicularly instead of along the box. Virgin cork can easily be broken with the hand, and it is rarely necessary to cut it.

The lighter colours of the edges here and there may be regarded as a blemish, but the newness will very soon be toned down by exposure, so that it is not necessary to colour them artificially. None of the pieces should be too large, for though it may be easier to nail down one big piece, the appearance of several

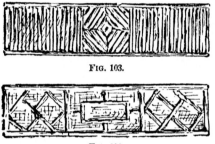

FIG. 103.

FIG. 104.

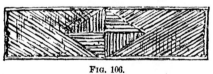

FIG. 105.

FIG. 106.
DESIGNS FOR BOX FRONTS.

of comparatively small size will well repay the extra labour of nailing down. Very often the cork is cut level with the top and bottom of the box. This may very easily be managed by sawing off the rugged ends after the pieces are nailed down; but it can hardly be considered an improvement, as the slight projections above and below seem more in harmony with the general character of the cork front than straight edges. Of course, if the box does not project beyond the ledge in front, the cork must be levelled at the bottom.

The cork may either be left in its natural colour or be decorated by painting, or rather touching here and there with bronze powders mixed with a little varnish. When judiciously applied, the metallic sheen of these powders in various shades of colour has a charming effect. There is, however, a danger of overdoing it, and it should be said that the freshness of the powders soon wears off. The principal objection to their use is, perhaps, that they give an artificial appearance to the cork, but this is a matter which may best be left to the taste of the maker. Plain varnish, of which the cheap and commonest kinds do very well, is often considered to improve the appearance by brightening up the otherwise dead surface of the cork.

Another very effective means of decorating window-boxes may be found in pieces of sticks or branches in their natural state as cut from the tree. They are cut to suitable lengths, and then, after being split, are simply nailed on to the box. Various simple patterns can be arranged without much difficulty as suggested by Figs. 103 to 106, which speak for themselves.

A similar but stiffer effect to that of the pattern shown by Fig. 105 can be got by the use of thin, flat pieces of wood of uniform thickness and width bradded on, the appearance being much like that of an overlaid fret. In such treatment it looks better to let the outer pieces be rather more massive than the others, to which they form a kind of frame or border. Instead of pieces of wood such as spoken of for either of these two schemes of decoration, pieces of bamboo may be used, either alone or in conjunction with ordinary cane. The bamboo or cane should be split in two, the flat sides being fixed next to the wood.

An almost endless variety of decorative device is opened up by the use of bamboo and cane, and it is by no means a difficult matter to give a semi-Japanesque style to the boxes by this means, especially if the wood is painted black and varnished before the pieces are nailed on. The edges and corners may be very neatly finished by cutting a kind of groove or rabbet in the bamboo, splitting a piece equal to about the quarter of its diameter away, as shown in Fig. 107. Except at the knots, bamboo splits readily, and at these parts it can

FIG. 107.—BAMBOO CUT FOR EDGES AND CORNERS.

easily be sawn or cut. Novices may be reminded that the edges at the surface of bamboo, being very fine and hard, cut the flesh like a knife, though there is no danger if ordinary care be used.

An appearance very much resembling that of a tiled box may be got by covering the fronts with pieces of floorcloth or linoleum, a "tile" pattern being chosen. Owing to damp, these materials are apt to stretch, so that they do not lie flat and close to the wood. This defect may be greatly lessened by fastening down with ordinary linoleum cement, a tin of which is quite sufficient to do for several boxes. It cannot, however, be efficiently applied to damp wood, so that it is almost impossible to make the cloth adhere properly to boxes which are in use. Common cycle-tyre cement is also an excellent adhesive medium, though it is not so easily used as the linoleum kind. Whatever it is, the cement may be used all over or only applied in patches. In the latter case the covering material will not remain so flat, the liability to bulge depending on the distance between which the cement is placed. A few brads should be driven in at the edges. Those who do not use any cement may fasten the floorcloth with small nails or tacks. If plenty of these are used, the material will not swell out from the wood to any appreciable extent. The resemblance to tiles may be still further increased by putting strips of wood on the front at the ends and top and bottom edges. Of course, neither floorcloth nor linoleum,

even if put on new, will long retain its gloss when exposed to weather. A coat of varnish, however, applied now and then, will remedy any dulness. To a casual observer the difference between encaustic tiles and the imitation would scarcely be distinguishable, and this cheap and effective means of decoration may be safely recommended to those who prefer the appearance of tile boxes.

Tiles are made in various sizes, but those most commonly met with are 6in. and 8in. squares. As there is more demand for the former than the latter, a much greater range of patterns is obtainable in them, and they are also very much cheaper, being sold at prices ranging from about 3s. per dozen upwards. Very good tiles for the purpose are to be had for about 6d. each, the price depending not so much on the quality of the tiles themselves, as on the quality and colourings of the imprint. Hand-painted tiles are of course out of the question for the purpose, on account of their higher value. In choosing tiles it should not be forgotten that, when placed on a window-box, they are not likely to be subjected to a close inspection, so that a good bold pattern is to be preferred to one with a quantity of minute detail. Remembering that the tiles are of a fixed size, the box should be of a length to take a certain number, leaving a margin over of, say, 1in. at each end. Thus, if we want a box of somewhere between 3ft. 6in. and 4ft. long, seven 6in. tiles will be used, giving 3ft. 6in., to which, by adding, say, 2in., we arrive at the required length, 3ft. 8in., to which the box is to be made. Tiles can be cut; but, in most instances, the uniformity of the pattern is destroyed, and it will be better to avoid the necessity for doing so by a little arrangement beforehand. Such a box as that already described will do very well for finishing with tiles. These cannot be stuck on; but must be kept in place by a rabbeted moulding along the top and bottom, where, by the way, a similar margin to that at the ends should be left. The margin may be of any width, the only requisite being that it is wide enough to afford a sufficient support to the moulding, and consequently to the tiles. It is quite possible to dispense with an ordinary rabbet and still to hold the tiles in.

F

Some pieces of wood the same thickness as the tiles will be wanted. These must be cut into strips the width of the margins already referred to. A neater job will result if these pieces are a trifle wider than is actually necessary, so that they may be afterwards planed down level with the top edge and bottom. Four pieces will be required for the front of each box, so that a sort of frame within which the tiles will fit is formed. Four similar pieces, of which the thickness may be the same as the others, but not necessarily so, will also have to be prepared. These must be a little wider than the others—say about ¼in.—so that

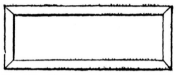

FIG. 108.—MITRED FRAME OF FALSE RABBET.

they overhang the tiles to this extent, and when nailed down hold them securely. This recess binding the edges of the tiles is virtually a rabbet. The four facing or front pieces should be mitred at the corners, as shown in Fig. 108, especially if the edges are rounded

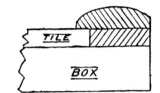

FIG. 109.—SECTION SHOWING FALSE RABBET AND TILE.

off, as they may easily be. They are represented so finished in Fig. 109.

No support is required between the tiles, the hold at the top and bottom being sufficient. The best way to proceed with the work will be to put the moulding on at top, bottom, and one end. The tiles can then be pushed in from the other, and the moulding nailed down afterwards.

By giving the boxes inside a coat of tar or melted pitch, either of which is easily applied, the durability of the wood will be much improved.

A TOWEL-HORSE.

THE towel horse (Fig. 110) here described measures 2ft. 6in. extreme height, and 2ft. 4in. long; the uprights (A) are 2in. by 1in. stuff, and cut 2ft. 2in. long, which allows 1in. for mortising into the feet. The crossbars (B) for rails are 8in. long and 2in. by 1in., the same as the uprights, and are let into the uprights at right angles, the crossbars being cut out as shown in Fig. 111; a similar cut being made in the upright, they will fit together flush, and will only require glueing to fix them. The feet are 8in. extreme length, 3in. deep, and 1in. wide, and into them uprights are mortised; the underparts are hollowed out, and top corners taken off. The rails can be either round or oblong, with the edges chamfered off and let in at the ends half the thickness of the crossbars and uprights; care must be taken to

FIG. 110.—TOWEL-HORSE.

make them fit tight, as they are only glued afterwards. The edges of the cross-

bars, uprights, and feet might be chamfered and the ends rounded, should the towel-horse be thought too heavy in appearance.

Towel-horses, as a rule, have turned uprights and crossbars in various patterns, but we give this simple pattern

FIG. 111.—CROSSBAR OF TOWEL-HORSE.

supposing the amateur does not possess a lathe, or, in the event of his having one, that it has not sufficient length of bed to allow of turning the required length. The pattern shown, when carefully made, has a very neat appearance.

A PEDESTAL CUPBOARD.

THIS cupboard (as shown in Figs. 112 and 113) should be of the same wood as the remainder of the bedroom suite, but yellow pine has a good appearance, is cheap, and is easily worked.

The design here given is very plain compared to many one sees at furniture warehouses, but it is no worse for that.

The pedestal is 1ft. 9in. high, without the top, 14in. wide, 12in. deep, without the door. The two sides and door are panelled, and it is as well to get these out first, the door being 3in. shorter than the sides. Wood ⅝in. thick will do for the frames of sides and door; the stiles, or uprights of sides, being 2½in. wide, the top rail the same width, the bottom rail 3in. The rails are then morticed into the stiles about 1¼in. On the inside edges of the frame

run a ¼in. bead; this is of course done before putting the frame together.

The bead should not run the whole length of the stiles, although it will be necessary, when cutting it, to work from end to end; it must then be cut off, at a mitre angle, where the rails meet the stiles, and this must be allowed for when cutting the lengths of rails. For example: The sides are 12in. wide, the uprights 2½in. wide each; the rails will therefore be 7in. long when fitted, but on account of the ¼in. bead an extra ½in. must be allowed in the length of the rails when the beads are cut off at each end of the stiles where the rails fit in. This will apply equally to the door.

Before glueing the frame together, grooves must be made on the inside edges for the panel to fit in. If the amateur does not possess a grooving-plane or plough, a simple way is to cut the grooves out by means of a marking-gauge. Set the gauge so that when marking from either side it leaves a ¼in. space in the centre of the edge;

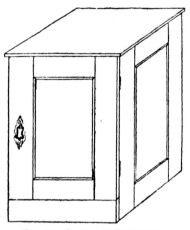

FIG. 112.—PEDESTAL CUPBOARD.

then, by exposing the cutter about ¼in. and filing it flat, it will cut two lines, ¼in. deep; the centre can then be taken out with a ¼in. chisel, leaving a clean groove. The panel should be ¼in. or ⅜in. thick, and if the latter the edges on all four sides must be tapered to fit in to the ¼in. grooves. The door is to be made in like manner, except that the beading on the frame should be inside as well as out for appearance sake when the door is open; the width of the door is 14in. The top measures 15in. by 13½in., back to front; this will allow of a ½in. overlap at the front and sides. The front plinth (Fig. 113, A) is 3in. wide, the same thickness as the door, and is screwed on to the front of the sides. The lower shelf (B) should be fixed on a level with the top of the plinth, resting on two fillets and fixed with screws from the sides. The shelf (C) must rest on fillets fixed to the sides, and the top is best fixed by screwing fillets on the under-side, so that the sides of the pedestal can be screwed

to them, which will leave the top surface free from screw-holes. The back need only be of thin deal fixed horizontally on to the edge of the sides.

In putting the cupboard together, first screw the fillets on the sides for the shelves, the centre shelf being 9in. from the top; screw on the front plinth and the shelf B to the sides, then fix the top, and, lastly, the back, the centre shelf to fit in without fixing. The door can be fixed either right or left hand as may be most convenient, with a pair of 2in. brass hinges which should be let in to the front edge of the sides, so that when shut they are flush with the edge. The handle and catch can be obtained at any ironmonger's, being made expressly for pedestal cupboards. In fixing the pedestal together, see that the heads of the screws are well countersunk in the wood; then fill in with putty, and size and varnish the whole work, or enamel according to taste. The top of the pedestal can be

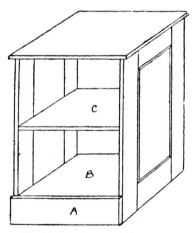

FIG. 113.—INTERIOR OF PEDESTAL CUPBOARD.

made more finished by running a moulding round it or by simply bevelling the edges on the top and underneath.

A WOODEN COAL-BOX.

FIG. 114 shows a coal-box of very simple design; but it will be as well for the amateur to carry this out in every detail before attempting anything more elaborate. It can of course be made of any wood, though the one the drawing is taken from is made of yellow pine, stained oak colour, and varnished.

Yellow pine has one great advantage, viz., it can be easily obtained as wide as 16in., consequently the sides are in one piece; whereas if ordinary deal is used, it will be necessary to have a joint

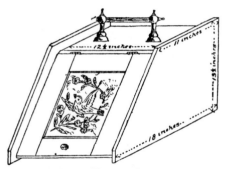

FIG. 114.—WOODEN COAL-BOX.

(the grain of the wood running horizontally), which should be avoided if possible; and, further, pine can be had entirely free from knots, and so makes a better imitation of walnut, of which it has the appearance when stained.

The measurements here given are for a medium size. The two sides measure 11in. at the top, 18in. at the bottom, and 13½in. in height, cut out of ½in. wood. It is better to cut the sides out first 18in. by 13½in.; then, when planed and the edges properly squared, cut off the front at the required angle, which is done by starting at the top 11in. from the back to within 1½in. of the bottom of the front edge. After cutting out

one side, place it on the second one to mark that out from, and then both sides will be exactly alike. The back, also of ½in. wood, is fixed between the sides flush with the back, leaving 1in. clear at the bottom and 1¾in. at the top. The top board should be ¾in. thick, and in it the handle is fixed. It is important that the top should be strong, for in the event of the box being carried about, the whole weight will be upon this part, which is fixed to the top of the back and to the sides, leaving a 1in. margin of the sides showing above it.

The most simple method of fixing all parts is, of course, by screwing. Nails must not be used for any part; screws will draw the top and back close to the sides providing the edges are perfectly square, and if the heads are well countersunk and puttied over they will not be noticed.

The bottom, ½in. thick, is better made of deal; it is not seen, and is stronger than pine. This, too, is screwed to the sides and to the bottom of the back, and will leave ½in. margin underneath, so that the box will stand on the sides. The bottom should be 14in. from the back, with a slip of wood, 2½in. wide, joined to it at a slight angle; this being done to prevent the coal falling out should any get spilt inside.

A wooden coal-box must of necessity have an inside lining of iron made the same shape as the sides, and to fit loosely, say ½in. margin all round, in order that it may be taken out easily for filling. The lining is usually beyond an amateur's capabilities, and should be entrusted to an ironmonger. It should have two handles of iron-wire fixed on the sides at front and made to fold inwards when not in use, and the whole should be of galvanised iron, as it will not rust should the coal be damp at any time.

The top of the coal-box will measure 10in. from front to back, the front edge bevelled to the same angle as the front of the sides.

The door is the most important piece of work. It consists of a frame 3in. wide all round, the centre panel being brass relieved with repoussé work; or it can be an ordinary glazed tile with a pattern; or a plain tile to be hand-painted; or, again, a plain panel of wood—this resting entirely with the amateur. The brass panel is very effective, and can be obtained, with the handle and knob, through any ironmonger, and in various patterns. The width of the frame will therefore depend upon what is decided upon as a centre. The width of the coal-box between the sides is 12½in. The door will be 12½in. square, and made of ⅜in. wood. The top and bottom of the door are mortised into the sides, the tenons being 1½in. long. The inside of the frame must have a rebate on its four sides ⅜in., and the depth of the re-bate will depend upon what the panel is. Should wood be selected as the centre, let it be of ¼in., the front surface slightly bevelled, say 1¼in., level on the four sides. If a tile or brass plate is used, it will require backing with ¼in. wood, not only to keep the panel firm, but to present a finished appearance when the door is opened. Brass hinges 2in. long should be used to hang the door to the top, and be let in to the top of the door, so that when they are closed they will be flush with its edge. The door when finished will rest on the small slip fixed at a slight angle at the bottom, and will allow an equal margin of the sides and top to show.

The scoop can be obtained at the same place as the handle and panel, with either wood or china handle, and a band 1in. wide screwed on to the back of the box for the scoop to fit into. The band can be made by the amateur, and may be cut out of sheet brass or sheet iron, the latter being blacked over when finished.

A DOUBLE WASHSTAND.

FIG. 115 shows a double washstand with tiled back that can be made at a very moderate cost and with little trouble. Made in yellow pine and sized and varnished, it makes an effective piece of furniture; or made in ordinary yellow deal, and afterwards primed and enamelled, it is equally satisfactory. In both stands ordinary deal quartering (2in. by 2in.) is used for the legs; but care must be taken to select wood as free from knots as possible, particularly when yellow pine is used for the other parts, as the latter is or should be nearly free from knots, so that when varnished the difference in the two woods may not be noticeable. When you have selected suitable wood of the size given for the legs, plane it up square, and cut the four legs the required length, viz., 2ft. 6in. These should then be tapered, starting about 9in. from the top, and reducing gradually from 2in. by 2in. at that point, to 1¼in. by 1¼in. at the bottom. The easiest way is to mark the required taper on the wood, and then saw outside the pencil line, so as to allow for finishing with the plane. The double stand measures 3ft. 9in. long, including 2in. overlapping at either end, and 1ft. 8in. back to front.

The frame must now be made. The side-pieces marked A are of pine, 3½in. wide by ⅜in. thick, are mortised into the front and back legs, and measure 1ft. 3in. when fixed. The back of the frame can be of ordinary deal, but rather wider, say 6in., as it will make the legs more rigid, and will allow of a tenon 5in. wide, the tenons on the side-pieces being 2½in. Care must be taken to make the tenons fit tight everywhere, for, with the exception of the front frame where the drawers are, the firmness of the stand mainly depends on these.

Next comes the front : the bottom of the front frame for the drawers is mortised into the front legs, and should be 3ft. 1in. long (when fitted), 6in. wide, and ¾in. thick; make the tenon 1in. wide, and as long as possible; the top of the frame the same size, but dovetailed into the top of the legs; the partition between the drawers should be of 1in. wood, 3in. high, and can be screwed or

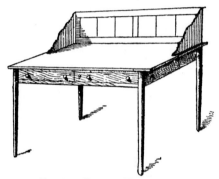

FIG. 115.—DOUBLE WASHSTAND.

mortised into the top and bottom of the frame.

This being done, the interior of the frame must be made for the drawers to slide on, and also to give additional strength to the stand. In the centre, fix a piece of deal 4in. wide and ¾in. thick from back to front. This can be easily done by two long screws from the back, and at the front by splicing; that is to say, screw a small piece of wood 4in. square under the bottom of the front frame on the inside edge with 2in. overlapping, to which screw the centre runner, which must be exactly flush with the bottom of frame. The side-pieces of the interior can be fixed in like manner, with the addition that if needed they can also be screwed from the side of the frame, the screw-holes being well countersunk and puttied after screwing, so as to leave no sign of the screws.

These internal side-pieces should be of the same width and thickness as the centre one, and will want a small piece cut out at the back-end where the legs come. These internal pieces or runners can be grooved ¼in. with a grooving-plane or plough—the centre one on both sides and the end ones on one side, so that thin bottom panels of ⅜in. deal, bevelled down to ¼in. at the edges, can slide in; but this is not necessary.

The drawers should next be made, and they will measure about 1ft. 6in. wide, 3in. deep (outside measure), and 1ft. 6in. back to front. The front and sides should be of pine, the back and bottom of deal; the front being ¾in. thick, sides and back ⅜in., and bottom ¼in. or ⅜in. The sides must be dovetailed flush with the ends of the front, the length of the dovetail not to be more than three-fourths of the thickness of the front, so as not to show when the drawers are shut; while the back of the drawers may be nailed, or better still dovetailed, to the sides, when thin slips, ¼in. square, must be fixed with glue and brads to the bottom of the sides, and the bottom of the drawer laid on these, similar pieces being fixed on the top. The drawers must run on the sides in any case, and where the thickness of the wood will not allow of grooving, the above is a simple method of fixing the bottoms. Slips of wood must be screwed on to the interior of the frame close up to the sides of the drawers, leaving sufficient room for them to slide in and out easily.

The top comes next, and as pine can be had from 9in. to 16in. wide it will only require one join to make the top of the required width, viz., 1ft. 8in. This will take two boards, 3ft. 9in. long, 10in. wide, by ¾in. thick. Shoot the edges perfectly true with the trying-plane, then glue and clamp them; or, better still, groove the two inside edges ⅜in. deep, ¼in. groove, and insert a thin slip, the grain running at right angles to the boards; then glue and clamp. The front and side edges of the top are left square, or can be bevelled if desired, or a moulding run round, if the amateur possesses the necessary planes. This is purely optional. The back frame for the tiles is made out of ¾in. wood, 1¾in. wide, with a division in the centre of the same width ; the back is rebated ¼in., to receive the tiles, 6in. by 6in., and the rebate must be of sufficient depth to allow of the tiles laying in flush with the back. The inside edges of the frame can be chamfered, which gives the whole thing a more finished appearance. Six tiles are required for a double washstand—three each side of the centre piece of wood—pattern and colour to suit the amateur. They are kept in their place by a thin board, screwed on at the back, rather under the size of the frame, so as not to show.

The side-pieces or wings (Fig. 116) are 9in. high, 10in. long, and ¾in. thick, and

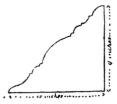

FIG. 116.—SIDE-PIECE OF WASHSTAND.

are screwed on to the side of the frame. The pattern should first be drawn to size on stiff brown paper, then cut out and placed on the wood, so that both wings will be exactly alike. They can then be cut out with a frame-saw or fret-saw, and finished off with glass-paper. On the top of the frame a slip of wood 1¼in. wide, either half-round or bevelled edges, gives a finish to the whole thing, and can be bradded on.

The top of the stand can now be put on, either loosely with small pieces of wood screwed on underneath to prevent its slipping either side or forward, or it can be permanently screwed on from the top, well counter-sinking the screws, and puttying the holes afterwards. The former plan is preferable, as it is handy to be able to move the top at times. The top should overlap 2in. either side and 1in. at the front. The tiled back can be fixed from the rear by two pieces of wood partly screwed both on it and on the back legs, which will keep it firm.

This will complete the stand. Now finish off with sand-paper ; size and sand-paper it again to get a perfectly smooth surface, then varnish with best copal varnish, and after one coat of varnish is well set, sand-paper again, and put on a second coat of varnish. This will make your washstand perfectly waterproof, besides giving the whole thing a much better appearance.

Should it be more desirable to make the washstand in ordinary deal, it should be primed when finished, and afterwards one coat of ordinary paint (any ground colour) applied ; this when dry should be well sand-papered ready to receive the enamel paint, which in French grey, white, or pale pink, is very effective.

A COMBINATION BEDROOM SUITE.

THE illustration (Fig. 117) shows a combination bedroom suite, suitable for a small room, where space is limited, as it has the advantage of being a chest of drawers, washstand, dressing-table, and towel-horse combined ; and it may be here mentioned that the towel-rail shown separately (Fig. 120) can be screwed on to either side, according to where the piece of furniture may be placed. The one here described was made for a very small room (10ft. by 7ft.), consequently the size was reduced as much as possible—to only 2ft. 9in.

long over all ; still, the drawers were of a very useful size. Of course, for anyone who may feel disposed to make a similar article it is not necessary to follow the measurements as regards the length, while the height and width may remain the same.

The choice of wood with which to make it must remain an open question : any wood, from common deal, upwards. The one shown was made in yellow deal, and simply varnished afterwards. This wood, if well selected, shows a very good grain, and when varnished is an admirable colour for bedroom furniture.

The frame (Fig. 118) measures 2ft. 6in. long inside (*i.e.*, between the uprights), 2ft. 9in. high, 1ft. 8in. back to front. First build up the front of frame. The uprights (A) are 4in. wide, the top rail (B) same width, the other rails 3in. wide. The top rail is dovetailed on to the uprights, the other rails mortised into them (as shown), all edges being flush at front. The middle partition (C) is 4in. wide, 9in. long, which is the space between the top and second rail. The frame should be of 1in. stuff, which is practically ⅞in. when planed, and the front of frame should be of the same wood as will be used for all parts showing ; the parts of the frame not seen may be of ordinary deal.

The front edges of the frame can be left square, or a ¼in. bead may run round the inside edges of the uprights, top and bottom rails, and both edges of the middle rails and partition. This is a great improvement, and if the amateur has not a bead-plane, he should get a ¼in. one (the price is about 2s. 9d.) ; he will find it useful in many ways. Next prepare the back uprights and side rails ; the former will be the same height and width as the front uprights. The side rails are 1ft. 5in. long, 4in. wide ; in order to fix them cut a slot 1in. deep in the front uprights, so that the side rails fit close up to them, and see that the surface is perfectly even, otherwise it will affect the sliding of the drawers ; the rails must fit tightly

in the slots and be glued ; the back of the rail must be screwed to the uprights (H), and the rails notched out, so that the outer edge comes flush with the uprights.

The middle rail (D) is the same length and width ; the front being screwed underneath to the partition (C), the other end to the middle upright.

The partition between the top drawers is to be mortised into the top and second front rail before the other parts are put

FIG. 117.—COMBINATION BEDROOM SUITE.

together. Glue the front rails into the uprights, and use only screws to fix the back uprights to the side rails. Fill up the back with ½in. matchboard fixed horizontally. This will materially strengthen the frame, and, being tongued and grooved, prevents any dust from getting inside the drawers from the back.

The frame being finished, the next thing is to make the outer case, which must not be done hurriedly, as so much depends upon the good workmanship. Wood ⅝in. wide for the sides, and ¾in. for the top should be used. In consequence of the width, it will be necessary to have one joint in each side piece and the top. Cut the boards for the sides 2ft. 9in. long, being careful to plane the edges true, then glue and cramp each side piece together. If the glued joints are sound—*i.e.*, the edges joined being perfectly true so that daylight cannot be seen between them—they will be found quite firm enough, it being unnecessary to make grooved joints for the sides, though it is certainly better for the joints of the top board to be tongued as well as glued, as this acts as the washstand, and consequently water is very apt to be spilt, which might in time loosen a glued joint and drip into the drawers beneath. Having jointed up the side boards, see that the edges are squared true, then fix them on to the frame by screwing from inside, being careful that the point of the screw does not appear through. The top is to be fixed on in like manner, and should overlap the front and sides 1in. each way, and a small moulding be run round the front and sides ; or if the necessary plane be not forthcoming, bevel the edges equally on each side, so that they form a true mitre at the front corners. It is as well to leave the top unfixed until the last, and a good plan is to screw small blocks of wood underneath it, and in such positions that the top can be put in its place and be prevented from slipping in any direction.

Now prepare the arms to support the looking-glass. These are 1ft. 9in. extreme length without tenon (Fig. 119), and 4in. wide at the base ; the two lines represent the section of shelf, which is fixed 4½in. from the bottom. To fix these, leave a tenon at the bottom, and let them into the top board, the back to come flush with the back of the top, and to be fixed at a distance of 18½in. between them. The width of the frame of the glass being 18in., this will allow of a ¼in. play each side. See that the tenons fit tight, and glue as well, and they will be kept firm by the shelf as shown in the drawing.

The side-wings (A, Fig. 117) are 10in. long, 6in. high, and are of ½in. stuff. The shape is simple enough, and can be cut out with a frame-saw. Of course this need not be adhered to ; any pattern in character with the whole piece of furniture will do. These are fixed a slight distance from the edge of the sides, and screwed from underneath. The shelf, the whole length of the back, is 3½in. wide, of ½in. stuff ; the front edge is chamfered off; then the back need only be of very thin wood, 6in. wide, and screwed or bradded on.

The frame for the looking-glass measures 22in. by 18in., frame 1¼in.

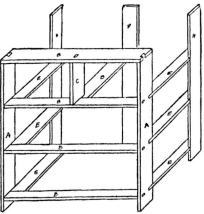

FIG. 118.—FRAMEWORK FOR COMBINATION BEDROOM SUITE.

wide, cut out of wood 1in. thick, with rebate quite ½in. deep. The front surface can then be slightly rounded, as in the case of the ordinary looking-glass frame, or fluted by means of a " bead-router," the corners are mitred, and the back filled up with a thin board.

The plinth round the front and sides at the bottom is about 2½in. wide, chamfered at the top edge ⅝in. thick, mitred at the corners, and either screwed or nailed on ; in either case the heads to be well sunk in and puttied over.

The drawers will now complete the article. The fronts of these should be ¾in. thick, the backs and sides ½in. Cut out the fronts first and make them fit well in their respective places; so much depends upon a drawer fitting well. Of course all parts must be dovetailed; fix the bottom in grooves made in the sides, or resting on slips of wood nailed to the sides; and it is worth repeating, the drawers must run on the

FIG. 119.—SUPPORT FOR LOOKING-GLASS.

bottom edges of the sides, so fix the bottom of the drawer itself clear of them.

Fancy handles can be obtained at most ironmonger's at various prices and patterns; also the glass movements for holding the swing-glass, and these are far preferable to the glass-screws with wood knobs used in ordinary looking-glasses, glass movements being found generally in better work.

If the drawers fit tight, many advocate a free use of blacklead. It is better to make the drawers fit easily by good workmanship than to have recourse to blacklead.

Runners are to be screwed on side rails and middle rail of the frame to keep the drawers in position. Do not put

blocks at the back of the drawers, but make them as deep back to front as the frame will admit.

Well sand-paper all parts with coarse and fine paper. Size with clean size, and, when hard, sand-paper again, finally putting on two good coats of oak copal varnish, allowing the first coat to thoroughly harden before applying the second, and removing any rough places left after the first coat with sand-paper. If made in yellow pine, the varnish throws up the grain, and has a very good appearance. If the amateur intends enamelling his furniture, there is no need to use yellow pine at all, but all deal, the former being about twice the price of the latter.

The towel-rail, as shown in Fig. 120, can be fixed on either side. The brackets are 12in. high, projecting 8in. for the top rail and 4in. for the bottom rail, $\frac{3}{4}$in. stuff being used; the rails also being

FIG. 120.—TOWEL-RAIL.

$\frac{3}{4}$in. diameter and 14in. long, let into the brackets about $\frac{1}{4}$in. either end, and screwed from the outside. To fix the towel-rail, use small brass glass plates.

IMPROVING A MANTEL-PIECE.

DOUBTLESS there are many who occupy a house where the mantel-pieces are not the most artistic or of the best quality. The one that the present design was made for is of the commonest white marble: the shelf had been previously hidden by a mantel-board covered with crimson cloth having a deep border to hide as much of the front as possible ; and when no fire was required, artistic curtains covered the whole front. But in the winter months it is hardly safe to have drapery so near the fire even if the curtains are drawn back, and consequently the unsightly jambs would be as much exposed as without them. The design, therefore, as shown in Fig. 121, is made to illustrate what may be termed false jambs, which are made to fit over the existing ones. It is of course impossible to give precise measurements, as the fronts must be made according to the size of the jambs over which they are to fit. In the event of the mantel-piece having pieces projecting from the face of the jambs to any extent, the design as given will be of no use ; it is only of use where common mouldings are on the jambs and which do not project more than 1in. ; otherwise the false jambs would project too far into the room.

In order to more easily describe how to make the fronts, the dimensions of some that have been made will be given, and they will not be far out in the general run of cheap mantel-pieces. The fronts should be of yellow pine, which will be found the best of cheap woods, being free from knots. In deal or pitch pine will be found a certain amount of resin, which would ooze out if the wood were exposed to heat for any length of time. The wood should be ½in. thick for the front and sides. The front measures 3ft. 6in. high and about 10in. wide, as it must extend beyond the present jambs ½in. each side, so that the sides can be fixed flush with the edge of the front. The side farthest from the fireplace is 3ft. 6in. high and about 3½in. wide, so that it fits close up to the wall, a small piece being cut out at bottom to allow for skirting-board. The side nearest the fireplace is about the same width, but will require cutting out at the top to admit of its fitting close to the cross slab of the mantel-piece marked A in Fig. 121. These sides should then be screwed (not nailed) to the front as shown in the section Fig. 122, care being taken that the heads of the screws are quite flush with the front. Before fixing the sides, the edges of the

front board should be beaded or ribbed by means of the bead-router ; the edges of the shelves should also be done in like manner and the brackets supporting the shelves in the curved part.

The shelves measure 7in. back to front, and the shape as shown looks well. The shelves should be screwed on from the back with fairly long screws, which will prevent the wood from buckling or getting out of the flat when exposed to the heat of the fire. The brackets should be also fixed from the back with two screws each and by one each from the brackets to the shelves as shown in Fig. 122. The brackets measure 6in. by 7in., the longer sides being cut the way of the grain as they

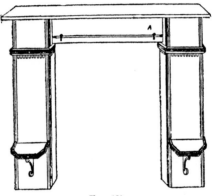

FIG. 121.

are screwed to the front board. The shelves should be fixed 12in. from the bottom ; this will give sufficient room for the fender or curb, or even fire-dogs to stand underneath them.

The moulding at the top should next be made and fixed. A full-size section is given in Fig. 123, which also shows a section of the tooth-moulding below it. The moulding is of wood, 1in. by ¾in., with a ½in. chamfer, leaving a thickness of ⅜in. at the bottom, as shown. The moulding is also screwed on from the back, 8in. from the top, the corners being mitred. There is no need to glue or otherwise fix the mitres, as the pieces of moulding at the sides being also screwed on they will draw up close to the front moulding, providing the mouldings are properly cut. The tooth-moulding below is formed out of several pieces of wood ⅜in. long, ⅜in. thick, ¼in. wide, the bottom edge slightly chamfered with the chisel ; these are fixed at distances of ¼in. apart by

FIG. 122.

means of needle-points, one in each being sufficient.

In the course of construction, it will be as well to fit the fronts on to the marble or stone jambs more than once, to see that they fit well and tight underneath the mantel-shelf, for if they do it will not be necessary to fix them in any other way. The amateur is also advised to measure both jambs separately—not taking the measurement of one as the measurement of the other—as in common mantel-pieces the jambs are unfortunately not always the same width, and, unless precautions be taken, a great waste of labour and materials may be entailed.

On the fronts of the false jambs, fix a thin beading down each side to cover the screw heads, this is ₁₆⁷in. wide by ¼in. thick, with beading on the front to match other parts, and is fixed with needle-points.

Next comes the question of finish, upon which opinions of course differ ; but the following is a good plan. Having well sand-papered the fronts, prepare about 2lb. of white-lead with turps and a small quantity of gold-size (about a wineglassful). Having mixed these well together, strain the paint through white muslin, or any material of like texture, so that all lumps or

surplus white-lead are left behind, then paint the fronts in the ordinary way. A second coat should be given in this way, allowing at least twelve hours for the first coat to dry properly, and should there appear any roughness, sand-paper again before putting on the second coat. The third and last coat should have more gold-size—about double the quantity to the same amount of white-lead—and of course less turps in proportion. To this add as much ultramarine-blue as would cover a threepenny-piece. This will make the paint a pure white, and the fronts look remarkably effective in this dead-white colour, while the paint not having any boiled oil in it will not blister with the heat. The fronts of course can be varnished if the amateur prefers a glossy surface ; but spirit varnish should be avoided.

The false mantel-board should have a border about 8in. deep, and this, with the false fronts, is to be preferred to any unsightly mantel-piece.

The fireplace for which the false jambs here described were made had an ordinary hearthstone, but a tiled hearth was added simply by laying the tiles on the hearthstone and fixing them with plaster of Paris. Portland cement makes a more lasting job, but should it be desired to remove the tiles at any time, the chances are they would break in the process of getting them up.

For a summer decoration, instead of having curtains fixed on thin rods underneath the mantel-board at the front, and drooping down so as to cover fireplace, fender, and all, the horizontal

FIG. 123.

slab on existing mantel-piece, marked A in Fig. 121, can be filled in with wood, painted to match the false jambs, and on it fixed a ⅜in. brass cased tube, with fancy ends, by means of small brass brackets, all obtainable at any ironmonger's, and on the rods put a dozen lin. cornice-pole rings. To these may be hung two small curtains (of any colour or material, but naturally something delicate in colour to harmonise with the jambs), which only cover the stove, and therefore leave the jambs, tiled hearth, and its furniture exposed. On the shelves at either side Art-coloured pots with ferns, or any plant to suit the fancy can be placed, and the whole makes not only a novel but also a very pretty summer decoration for a fire-place.

In character with the above, is an over-mantel in ordinary wood and painted to match the jambs.

A CHILD'S COT.

USEFUL, and at the same time not a difficult, article of furniture to make, is a child's cot; and the pattern described and illustrated in this article is one that is well within

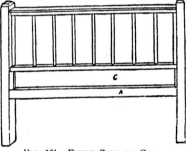

FIG. 124.—FIXED SIDE OF COT.

the capabilities of any amateur carpenter who can handle his tools at all.

The cot can be made of deal, or any other wood, though the former is quite strong enough and much cheaper. The cot here described measures 3ft. 9in. long by 2ft. 2in. wide between the posts. The posts are 3ft. high, and made out of 2in. by 2in. quartering, planed-up true, and the edges well chamfered or bevelled say to the width of ⅜in.; the tops in like manner. The bottom rails (A, Fig. 124) are of the same size, left

square, and mortised into the posts, so that the tenons of the side and end rail just meet at right angles in the posts. As the posts are all the same length either end can be used as the head of the cot; and this is of consequence, as will be seen further on. Fig. 124 shows one side of cot, which is a fixture. Figs. 125 and 126 show a part of the side which is movable, so that when the cot is placed at the edge of an ordinary bed, one side can be taken out if desired, and the cot can be placed at either edge with the movable side next to the bed.

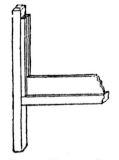

FIG. 125.—PART OF MOVABLE SIDE OF COT.

The bottom rails are fixed 12in. from the ground. The frame of the cot is shown in Fig. 127, and has a rail (B) 2in. by 2in. mortised into the side rails midway between the posts. Chair-webbing is then stretched across from rail to rail and interlaced, as shown in drawing, and tacked on to the top of the rails and middle rail also; this should be done before the sides and ends are made. The description of Fig. 124 applies equally to the ends as to the one

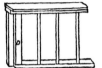

FIG. 126.—PART OF MOVABLE SIDE OF COT.

side. Therefore, taking Fig. 124: C is a board, 4in. wide, ½in. thick, which fits into grooves fixed on to the posts; these are of ½in. wood, project ¾in., and are the same length as the width of the board. Next is a piece of wood, 1¼in. wide, ¾in. thick, in which holes (⅜in.) are drilled, at distances of 5in. from centre to centre, and about three parts of the way through. These pieces are screwed to the boards and to the posts with long, thin screws, put diagonally. The upright bars are ⅝in. square, with the edges rounded off, and should be 15in. clear between the rails into which they are fixed. The top piece, into which the bars are also fixed, is likewise 1¼in. by ¾in., and the holes are drilled

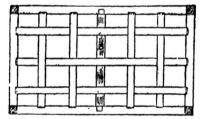

FIG. 127.—PLAN OF FRAME, WITH WEBBING FIXED.

right through. On the two sides of the posts are screwed pieces of wood ⅝in. square, 15in. long; the top rails rest on these and are screwed to them. The bars are then put in their places, and should fit into the holes tightly. The bars are made fully long, so that when

in position they can be cut off flush with the top. Finally a piece of wood, 1¼in. by ½in., is fixed on to the top, which will cover the bars, and are fixed with brads to the under piece, the brads being punched in and the holes puttied. Fig. 128 shows a section of a bar in position.

Thus far the two ends and one side are finished, and what remains to be done is only the movable side, as shown in Figs. 125 and 126. In this, the top and bottom rails into which the bars are fitted are the same width and thickness as the other side and ends, but the bottom rail is 1½in. shorter, to allow of a tongue ¾in. wide at both ends, which will make up the total length to 3ft. 9in. The tongues are screwed to the upright (D), which is 1¼in. by 1in.,

FIG. 128.—SECTION OF BAR IN POSITION.

and 15in. long. By referring to Fig. 126 it will be seen that the top rail is fixed flush with the outside edge of the tongue, so that when the side is put into position the rail will fit close to the posts.

The posts can be left as shown, but the amateur is recommended to buy a set of brass bedstead knobs and caps (2in. knobs), which cost about 1s. or 1s. 3d. the set. To fix these, four hard-wood pins, the size of the holes in the knobs, will be wanted. These should be fixed into the posts, say, 1in. deep, leaving 1½in. projecting, over which the caps are put and the knobs forced on. The knobs have a screw-thread in the holes; so if the pins are made fully large, the knobs will cut their way into them, and so make a firm job.

If the amateur wishes to have castors, those known as three-wheel wood-bowl castors, which cost about 1s. 6d. a set of four are recommended. These have a round iron plate, say 1½in. diameter, which is screwed on to the bottom of the posts, and the wooden roller or bowl is so fitted that it revolves on three small iron wheels.

The cot now being complete, it should be well sand-papered, and have a coat of priming, which is a mixture of white lead, turps, and boiled linseed oil, with a little dryers and red lead to harden it. This being done, a coat of enamel paint, or, better still, two coats, will finish the work. A delicate tint, say pale pink, or even pure white, looks well.

In making a cot, see that all the inside edges are bevelled or rounded off, and take care there are no splinters left that might hurt the little occupant.

A DUCHESSE DRESSING-TABLE.

THE dressing-table, as shown, measures 3ft. 6in. long, 1ft. 4in. wide, and stands 2ft. 6in. from the ground: in short, the same height as the washstand previously described. It is made

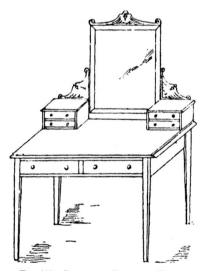

FIG. 129.—DUCHESSE DRESSING-TABLE.

also either in yellow pine or in deal enamelled. In any case the legs are cut out of deal quartering, 2in. by 2in., free from knots, and tapered to 1¼in. by 1¼in. at bottom. In every particular it is made precisely the same as the washstand, as far as the table is concerned. The top is of ⅞in. stuff joined in the centre, and overlaps the frame 1in. either end and front, so that it will really measure 3ft. 8in. long, 1ft. 5in. wide; the front and side edges are to be chamfered, rounded, or moulded as drawn, though the last form is certainly the best and most easily accomplished. In any case great care must be taken to have the plane in good order, for when planing the sides, which will be cross grain, the wood is likely to tear unless the tool is in good order. The top of the table must be made first, and joined and planed up true before the moulding is put on, as it will be found easier to cut the two boards fully long, so that when fitted together the ends can be squared and cut to the exact size. To give additional strength to the top, two strips of wood can be screwed on underneath from front to back. This may be considered extremely amateurish, but they are not seen, and should the wood not be well seasoned the strips will prevent warping, which would assuredly happen if the temperature of the bedroom was at any time higher than that in which the wood had been previously kept.

The table being finished, the next things to be made are the side drawers and cases on the top. They should be 10in. long, 5in. wide, and 5in. high without the top; the top should overlap ½in. at front and at one side, which will therefore measure 10½in. by 5½in.; while the front and outside end should be chamfered, rounded, or moulded to match the table. If moulded it must be less in proportion—⅝in. moulding for top of table, and ½in. for top of side drawers, wood ½in. thick to be used throughout.

FIG. 130.—SIDE OF CASE.

The sides of the cases should be cut as Fig. 130, and let in the table and the top; in the latter about ¼in. is sufficient, as the tenons must not appear through it, and when glued they will be sufficiently firm; the division between the drawers are to slide in the grooves, ¼in. deep, and be glued. The bottom simply fits in between the sides, and is fixed to the table from underneath. The tenons at the top of the side-pieces must fit nicely, as they will show at the inside ends of the cases; those at the bottom can be ¾in. long or more. The divisions between the drawers should be the same width as the sides. The backs, ³⁄₁₆in. or ¼in. thick, must be bradded on, after which the cases can be fixed in their places, the outside ends coming flush with the top of the moulding on the table.

The arms that support the frame of the looking-glass should be of 1in. stuff, and measure 9in. either way. The pattern should be first drawn and cut out on paper, in order to get both arms the same, and then the wood must be cut out by it with either frame or fret-saw. Should the amateur have the slightest knowledge of carving, and possess the necessary tools, a much better appearance can be given to the arms, if relieved as shown in the drawing. This also applies to the addition to the top of the looking-glass. The arms must be fixed very firmly to the small cases, and flush with the backs. As it would be found difficult to screw them on, and not at all satisfactory, the best method is to fix them by means of two

strips of wood at the back, half on the arm and half on the case; thin iron plates, however, are less conspicuous, and will give greater strength to the arms, as the glass is invariably hung at a slight angle when in use.

The frame of the looking-glass measures 2ft. 2in. high and 1ft. 9½in. wide; this will allow ¼in. space either side between the frame and the arms, so that the glass will swing quite clear, and when upright and in position a space of 2in. will be between the bottom of the frame and the table. The frame should be of 1in. stuff, 1¼in. wide, the face of it rounded slightly; the back of the frame should have a rebate ½in. and ⅜in. deep, or thereabouts, according to the thickness of the silvered glass. The corners of the frame are mitred, and can be cut perfectly true with the aid of a mitre-block. They should be glued and bradded from each side, and clamped together till the glue is perfectly hard, and this will be found sufficient to hold the frame together firmly, particularly as a back will be screwed on to protect the glass as well as for appearance sake. This back should be of ¼in. stuff, the edges chamfered, and it should not come quite to the edge of the frame. The frame is to be hung by what are known as glass movements, which are much preferable to the old-fashioned knobs and screws. These movements should be fixed exactly midway on the frame, and on to the arms about 1in. from the top. The ornamentation for the top of the frame should be cut out similar to the arms. This need be only ½in. stuff, and an easy method of fixing it is to run a ¼in. groove down the centre of the top of the frame, bevel the bottom edge of ornamentation to fit tightly into it, and then glue. The groove should be made before the frame is fitted together. If the silvered glass be bevelled, the rebate in the frame should not be more than ¼in., or so much of the bevel will be hidden. Bevelling adds considerably to the cost; but if this is not an object, it is well to add it. All these little details help to make a piece of furniture so much more effective, and give a better finish.

The dressing-table is now complete, and simply requires to be sized and varnished, or painted and enamelled, according to the wood employed.

G

Some amateurs may feel inclined to use woods other than those mentioned, for instance, mahogany or walnut. Either of these add very considerably to the cost, unless veneer is used ; but this is generally beyond the amateur. Then these woods would require to be French polished, which is a very difficult operation except to a professional polisher, and it takes away the charm not to make and finish the whole thing oneself. A very nice addition to a dressing-table are brass square socket-casters, 1½in. socket.

The drawers are not described, they having been given in a former chapter, the instructions in which apply equally to any size drawer.

A CHEST OF DRAWERS.

THIS chest of drawers is to be *en suite* with the washstand described at pp. 70-72, and therefore made either with yellow pine varnished or with ordinary deal enamelled; it will, however, be described as being made with the former.

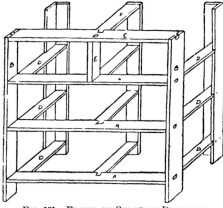

FIG. 131.—FRAME OF CHEST OF DRAWERS.

In making the frame (Fig. 131), the front is to be of pine, the remainder of ordinary deal, as when finished it is not seen. The frame measures 3ft. 3in. long. 3ft. 3in. high, and 1ft. 4in. front to back. 1in. stuff to be used through-out, though when planed up this will probably diminish to ⅞in.; if more convenient ¾in. wood will do, but 1in. is preferable. The front of the frame is to be made first : the uprights B and top rail C are to be 4in. wide, and the top rail is to be let into the uprights in the form of a dovetail. The drawer-rails A are to be 3in. wide, and let in the sides of the uprights with a tenon (1in. wide) 1in. from each edge. The drawer-rails must be fixed at certain intervals, *i.e.*, allowing a depth of 10in. for the top drawers, 11in. for the middle drawer, 12in. for the bottom drawer, and the bottom rail to be 3in. from the ground, to allow of a plinth being fixed at the front and sides, of which more hereafter. The division E between the small drawers is 4in. wide, and is let into the top and second rail, and will, therefore, project 1in. beyond the second rail inside.

The front edges of the frame can be left plain, but care must be taken to plane all edges perfectly true, so that when fitted the front edges will be properly flushed. Much depends on this, for when finished the sides will be fixed flush with the frame in front, and the two thicknesses of wood should appear almost as one. To give a more finished appearance, the front edges of the frame may be run round with a ¼in. bead, as also both sides of the second and third rails, division E, and the inside edges of the top and bottom rail and uprights. The uprights at the back (three) and back of sides are to be 3ft. 3in. long, 4in. wide, and the side pieces to be screwed or nailed to the outside uprights of the back.

The runners for the drawers must next be made, and all are to be 4in. wide and about 14in. long. The side runners must be let into the front uprights 1in., which will bring them flush up to the front rails—a nail or screw put in diagonally into the uprights will be sufficient to hold them in the front—and be nailed or screwed into both the back and side uprights at the back, cutting out a piece 4in. long and 1in. wide, to allow of fitting flush with the outside of the upright at the side of the back. The middle runners must be let into the top, third and bottom front rails in the form of a dovetail, which must be cut V-shape on the edge (Fig. 132) to fit into the front rails cut out in similar shape so that the runners fit flush with them; the other end of the middle runners must be fixed in the middle of the back upright, similar to the side ones. The middle runners must be cut at least 1in. longer than the side ones, in order to allow for the dovetail. The second middle runner, not hitherto mentioned, is the same length as the side ones, fitting up flush with the front rail, and screwed underneath to division E, which projects 1in. inside beyond the front rail in order to admit of this. It may, perhaps, be as

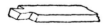

FIG. 132.—MIDDLE RUNNER.

well to mention that the frame must not be finally screwed or glued together until the whole has been fitted, as it will be found necessary to take out the

front rails (A) in order to fit in the dovetail of the middle runners (F).

The back of the chest of drawers should be covered with match-boarding (½in.), fixed horizontally, starting from the bottom, the ends to be cut flush with the side of the frame.

The frame for the drawers may seem

FIG. 133.—CHEST OF DRAWERS FINISHED.

a little complicated at first sight, but it is really extremely simple; and by making a frame and then the outer case, there is no fear of the chest when finished being anything but perfectly firm.

In making the sides and top (as in Fig. 133) the width will necessitate at least one joint, the fitting of which has already been described in the chapter on " A Washstand." ¾in. wood is thick enough for the sides, and 1in. for the top; or if ¾in. be used as well for the top, it will be better to screw two stretchers underneath the top, from back to front, about 4in. wide and 8in. from either end; these will not only strengthen the top, but serve also to keep the joints together.

The sides are fitted on first and screwed to the frame from the inside, the top to lap over about ¾in. each end and front, which must be allowed for when cutting out. The moulding run round the sides and front of the top is the usual OG, as it is termed, for which there is a regular plane; but it can be done with an ordinary rebate plane, which every amateur should possess just as much as a jack plane. If a rebate plane be used to make the moulding, which should be ⅜in. wide, first cut a rebate ¼in. deep, and then round the wood off gradually to the edge ; but great care must be taken to do this evenly, so that the mouldings shall be true when they meet at the corners.

The top must be screwed on similar to the sides from underneath.

To complete the chest, all that is now required is a plinth at the bottom to be fixed on the front and sides. This should be about 3in. wide, ⅜in. thick, with the top edge bevelled, and when fixed should just cover half the thickness of the bottom rail, so that the width of the plinth depends entirely upon the height of the bottom rail from the ground. The side plinths must be screwed on from the inside and mitred on to the front plinth. To do this accurately, a proper mitre-block—an inexpensive but useful addition to a workshop—must be used. To fix the front plinth, screw two small blocks underneath the bottom drawer-rail to which the plinth is to be fastened. This will have to be done from the outside, so should be nailed, as the heads can be punched in and the holes puttied so as to be practically unnoticeable. The drawers will require a good deal of attention, especially the long ones, to make them fit properly. The fronts (¾in. thick) should all be got out first and fitted, and the edges planed perfectly true—a frequent use of the square is never time lost, for, if proper care be taken at first, it comes almost as easy to do a thing right as wrong. The sides and backs of the drawers are ⅜in. thick, and the drawers must always slide on the sides, which want almost

as nicely fitting as the fronts. The sides must be grooved to receive the bottom, or brad and glue small slips on the sides to lay the bottom on, putting similar slips on the top to keep it in its place. The drawers should be dove-tailed, as described in the washstand chapter, and in the long drawers a stretcher should be fixed on the bottom, inside in the middle, from front to back, about 4in. wide, to prevent the bottom from splitting or bulging out in the event of heavy articles being placed in them. Slips of wood, 1in. square, must be fixed on the runners of the frame for the drawers to slide against, and stops at the back to make them fit in flush with the front will complete this useful piece of furniture.

Of course the worker is not bound to the measurements given, but this is a very useful and average size.

The chest of drawers being now finished, it must be treated in the same way as the washstand. If made of pine, well sand-papered, sized, and varnished—two coats, the first coat being sand-papered when hard before putting on the second. If made of deal, to be primed ; but should there be any knots they must be covered, first with knotting procured at any oil-shop, afterwards painted one coat of any ground colour paint ; and lastly, enamelled to match the washstand.

The handles or knobs for the drawers can be either of brass or wood ; in any case it is better to buy them. Brass drop handles, with fancy backs, either in stamped or cast brass, look well. There are many designs, which can be procured at any ironmonger's. Size, 3½in. for chest of drawers, and 3in. or 2½in. for washstand.

AN OVER-MANTEL.

THE over-mantel may be made in common deal—though yellow pine is preferable, so as to avoid knots as much as possible—and it may be stained and oiled; or painted to match the wood jambs of the mantel-piece.

The over-mantel here described measures 5ft. long and 3ft. 7in. high extreme, but the length will of course depend upon the mantel-shelf on which it will be placed; the height will stand good in any case. The length, without the wings A and B (Fig. 134), is 4ft., and height of front between the dotted lines 2ft. 4in.

The first thing to be done is to cut out the four uprights, marked C in Fig. 134, out of 1in. stuff, which, after planing

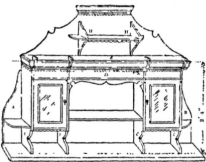

FIG. 134.—DESIGN FOR OVER-MANTEL.

both sides, will not measure more than ⅞in. The uprights are 2ft. 4in. high by 7in. wide, and a section of one end is given in Fig. 135, which shows the shape of the lower part. In order that the four uprights should be cut exactly alike, it is better to cut out the pattern in stiff brown paper or cardboard, and with it mark both sides of the upright, so that when cut out to the required pattern the edges will be square; if cut with a frame-saw, it will be found no easy matter, but requires great care, and a little trouble taken over this will

prevent a good deal of labour in finishing off afterwards. The two inner uprights that are joined to

FIG. 135.—SECTION OF END.

the centre shelf must be cut at the top, 4in. by ¾in.: this space will be filled up by what may be called the frieze (D, Fig. 134), which runs the whole length of the front, and will fit flush with the front edge of the outer uprights.

Having finished the uprights, it will be as well to prepare the frieze, 4in. wide and ¾in. thick, and as it will be fixed between the two outer uprights, it should measure 4ft. less the thickness of the two uprights; and, assuming they are ¾in. thick, this will bring the length of frieze to 3ft. 10½in.

The centre shelf is 2ft. long, and should be of the same width and thickness as the uprights. The shelf being 2ft. long, it leaves 12in. on each side, so that the spaces between the uprights on each side for the cupboards will be 12in. less the thickness of the uprights; in other words, the cupboards will be 10½in. wide, and this will determine the length of the bottom shelves of the cupboards.

First screw the inner uprights to the centre shelf, which should be 12in. from the bottom. The screws should be long, 2in. No. 10, or 2½in. No. 10 would be better, and the heads should be countersunk. This being done, fix the shelves of the cupboards to the two outer uprights and then join them on to the inner. To determine the height of the cupboard shelves it will be as well to mention that the height of the doors will be 15in., in order to fit close up to the frieze, and this being 4in. wide, the shelves should be fixed 19in. from the top; then fix the frieze with 1¼in.

screws, from the front to the inner uprights, and with 2in. screws from the outer uprights, taking care that it fits well and flush with the front edge of the latter.

It will no doubt be noticed that there is no silvered glass used in this over-mantel except in the cupboard-doors; but the design may be altered in this respect. Some may prefer having a glass fixed over the centre shelf, which can easily be accomplished by making a frame for the glass independent of the over-mantel, so that it can be put into its proper position after the over-mantel is fixed. And again, in place of the glass in the cupboard-doors, one might prefer a plain wooden panel with a hand-painted design. And further, the front edges of the uprights, the centre shelf, and the cupboard shelves, may be much improved by running a small bead on them with a bead-router.

The next part to prepare will be the moulding and the four pieces of wood which divide it, and which are fixed on the upper parts of the uprights. The moulding can be of any design; the one given in Fig. 136 is only to serve as a guide. It should be of wood 2in. by 1¼in., and made in one piece in order that the mitred corners may be true and well fitted. The design of the moulding, whatever it may be, should be followed when cutting out the four pieces to be fixed on the uprights, which are 6in. long, projecting 1in. above the moulding and ¼in. beyond the face of same (Fig.

136). The worker will do well to make a template for these four pieces, so as to get them exactly alike in shape and size, as in the case of the lower part of the uprights (C, Fig. 134).

It will be as well to proceed to fix the moulding and the four small pieces, first fixing the latter on the upright C by means of glue and one or two brads. The moulding, too, is fixed by the same means. The glue should be thin and not much of it used; rub the moulding on to the frieze to squeeze out any air-holes in the glue, and let the top moulding be a trifle higher than the frieze, the four small pieces extending 1in. above the moulding. The two corners of the moulding should be cut in a mitre-block, so that they will fit well and not show any join. The small wooden teeth, or dentils, underneath the moulding should be ¾in. long by ½in. wide by ⅜in. thick, and fixed with needle-points—one needle-point in each will be found sufficient.

At the bottom of the frieze, running the whole length of the front, is a thin half-round bead, ½in. wide. The fancy pieces over the centre shelf and under the cupboards are of ⅜in. wood, the centre piece being 3in. deep at the ends and 1¼in. at the top, the other pieces of sufficient depth to be level with the front edge of the uprights. They are cut out by means of a frame or keyhole saw, and should fit tightly, so that a little glue only will be necessary to keep them in their places.

The frames of the doors are 1¾in. wide, made out of ¾in. wood, rebated at the back to admit of the glass or panel, as the case may be, and if glass be used it should be bevelled. The frames are morticed together, the tenons not appearing through to the outside edges. The doors should be hung with brass hinges, 1½in. long, on to the outside uprights, and small brass drop handles with a catch on the inside to keep them fastened should be fitted, or in addition small brass locks that are known as "double-handed cupboard," in which event the keyhole should be covered with a small fancy escutcheon. These little matters all greatly tend to improve the general appearance of the over-mantel.

The over-mantel must have a top shelf (E) of not less than ½in. wood, fitting inside the frieze and uprights and resting on small blocks of wood. The wings A and B are 2ft. high (extreme), 6in. wide, and ½in. thick ; the small shelves of the same thickness,

FIG. 137.—BRASS GLASS-PLATES ON WING.

and are fixed to them on a level with the cupboard shelves by screws at the back. The wings are fixed to the over-mantel by means of small brass glass-plates (Fig. 137), 1in. long, obtained at any

ironmonger's, two on each wing and one on each shelf.

The upper part of the over-mantel is more easily made independently of the over-mantel itself ; when made it should rest on the top of the latter and also be fixed to the wall by means of glass plates, as in the case of ordinary wall looking-glasses. It measures 1ft. 3in. high and 4ft. long, and if made in yellow pine the back can be in one piece, as this wood can be obtained of this width, and will therefore save the trouble of making a join. Moulding of the same size as that already used is fixed along the top, 1ft. 6in. long.

The shelf brackets (H) are 11in. high, standing out 4in., which will be the width of the shelf, the centre and side shelves being fixed midway on the brackets ; ⅜in. wood being used throughout, and a fancy slip is glued underneath the centre shelf to match the other slips below. This will then complete the over-mantel, which only requires painting or staining as the case may be.

A SMALL WHATNOT.

LIKE every other piece of plain, simple furniture, the original idea can be worked up till finally a really handsome and valuable article is evolved. For this reason details will be given at length. Dimensions and such details can be varied according to requirements.

First consider the purpose of the article to be made, then arrange that the various parts shall be properly put together, in order that the whole may be sufficiently strong and durable. The size of the whatnot as shown is 3ft. 6in. high, by 1ft. 6in. wide, by 1ft. 4in. deep from back to front at the bottom to 6in. at the top.

For one of such dimensions as these the wood need not be thicker than ½in., and pine finished with an enamel paint looks very well. The fewer knots in the wood the better, and with a little care in the selection a good piece of red pine can be obtained almost free from them. If the stand is to hold books, knots in the shelves will not matter much, as they will be hidden. All the pieces must be made as smooth and even as possible by planing, scraping, and glass-papering, with ends and edges square and sharp, and, of course, the material must be thoroughly dry and well seasoned before fixing together.

The shaping of the front ends in Fig. 139

is very simple, and can be managed by anyone who has a fretsaw frame with adjustable clamps for the saw to face in any direction. Those who have only one of the commonest kind may, however, manage, with a little care, to cut

FIG. 138.—WHATNOT.

the rounded parts out. The easiest way will be to begin cutting at the narrow top end, working as follows: Lines having been drawn on the board as guides, or as it is called "set out" to the exact shape of the ends, and the necessary width having been got by gluing two or more pieces together, the back edge must first be got quite straight and true, then, by laying the square across, the top and bottom must be marked across and cut exactly at right angles. Now, beginning at the bottom, draw a line parallel with the back edge and 1ft. 4in. from it, unless the board happens to be exactly this width. The board must be made 16in. wide, and is set out from the

back edge, as it must ultimately be trued up; the front edge being cut off need not be shot true. At the top, mark off another line, but only 6in. from the back edge. Between this 6in. line and that giving 16in., or in other words the front edge of the board, there is now left a space of 10in. in width. Mark this off into four equal parts, always drawing the lines parallel with the back edge; it will be better to do this on the outside of the end. Next, on the same side, mark across the board lines indicating the position of the shelves, which are at graduated distances from each other, the bottom one being, say, 6in. from the bottom, and the top one about the same from the other end. The lines will, of course, be drawn by setting the square against the back edge, and neither they nor the others need be longer than to intersect each other (as shown in Fig. 140, which gives all the setting-out). The curves are set out with a pair of compasses, with the centre leg at a distance of about 3½in. above the shelf-line. As the space between each perpendicular line is

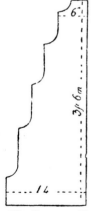

FIG. 139.—SHAPE OF ENDS.

2½in., this gives an interval or guard of 1in. at the front of the shelves, which will be much better than making them level with the "break." When this has been done, the actual work of shaping may be begun.

The wood being thin, those who have only a small back-edge saw in addition to a bow or fret-saw, will not experience much difficulty. The straight lines can be more quickly cut with a hand-saw. Cut along the outer line A till the curve is reached. Then remove the piece with a straight cut from the front edge. Never mind about the rounded part yet, but in sawing across be careful not to injure the edge above it, as this should want very little cleaning off afterwards. If the saw cuts into it, it will be more trouble than it is worth to hide the marks afterwards.

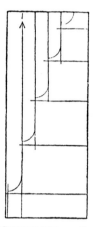

FIG. 140.—SETTING-OUT OF BOARDS FOR ENDS.

All the straight lines having been sawn, and the resulting narrow pieces removed, the fret-saw—the blades of which should be ot a good size—may be used to cut out the rounded portions. The ends, by having as much as possible removed before commencing with the fret-saw, will be much more easily handled, and allow the saw to be more freely worked. The edges will require cleaning up with glass-paper, and if the sawing has been irregular, a bull-nose plane will be an advantage on the straight parts, but not in the curves. While papering-up be careful not to round the edges, but to leave them nice and square.

The ends being ready, the shelves

may next claim attention, though, beyond saying that they must be smoothed and cleaned up with perfectly square corners, they require no special directions. The fronts may be rounded or bevelled; but as the shelves are not thick they will not look too heavy if left square. In cutting the shelves it must not be forgotten that on account of the back they must not be so wide as the ends are. If the thickness of the back is $\frac{1}{2}$in., the shelves will be necessarily that amount less, and it will look better if they are set back a trifle in front, say, to the extent of $\frac{1}{2}$in., than if quite flush. Particular care must be taken that the length of all the shelves is exactly the same. Now in the best class of cabinet work these ends would be either morticed or grooved, to hold the shelves, but for a painted article like this whatnot nails will do almost as well. The shelf-lines drawn across the ends show the position to insert them. Screws will be scarcely more effective than common nails driven in with the hammer. Holes should be bored for them through the ends with a bradawl, and the heads be well punched in. The bottom and top shelves should first be fastened, being careful to test the squareness of the job as the work proceeds, not only with them but also when fixing in the other shelves. The nails alone would probably hold the shelves securely, but little slips or blocks may be added underneath these. They must be firmly glued in. To prevent their showing, the small rails or brackets underneath the shelves are fixed.

FIG. 141.—BRACKET UNDER SHELVES.

The back must be exactly wide enough to fill the space behind the shelves, to which and to the ends it should be nailed. No object will be served by extending it down to the ground, as it will be quite sufficient if it be long enough to cover the back of the bottom shelf. In front a piece of wood must be fixed under this to enclose the opening. It should not be flush with the edge of the shelf, but be set back a trifle, and fastened as the other parts are with nails. A few blocks may also be put behind it if desired : these in nailed work are always advisable, as they serve

FIG. 142.—ALTERNATIVE SHAPING.

to stiffen the joints and prevent the strain on the nails.

The small pieces under the shelves may now be prepared and fixed. Besides hiding the blocks they strengthen the shelves, and prevent their bending down at the centre if used for heavy articles. The pieces may be prepared from some of the parts removed from the end board. Each may be made in one piece, but as they are straight for the greater part of their length, the easiest way will be to glue the small curved parts on. In width ½in. will be suitable. These are then glued into their places or nailed from below; the small curved ends are then fastened on with glue. Fig. 141 shows the connection between the parts.

A suggestion for alternative shaping of the end pieces may now be referred to, as well as one or two other little matters by which the appearance of the whatnot may be greatly improved without materially adding to the difficulties of construction.

As to the breaks in the ends, Fig. 142 shows an alternative. The round parts may be made with a large centre-bit, the holes being bored before the saw is used. As the bit will leave the under edges ragged if the hole were made from one side only, as soon as the centre point comes through below, reverse the board, and bore down to the hole already partially formed. By using the bit gently a good clean hole may thus be made, but it is indispensable that the cutter should be sharp. As a variety

from this alternative shaping, another is given in Fig. 143 which speaks for itself.

An improvement which will be of great advantage if the whatnot is to be used as a stand for ornaments and as a drawing-room piece of furniture, will be to line the inside with velvet, while a further improved appearance may be gained by fixing looking-glass at the back.

Now let us see about fixing the velvet and glass in an easy and at the same time presentable manner. To begin with the velvet should be stuck with glue or strong paste, rubbed on the wood, and that the parts of this to be covered need not be painted. A silk plush or velveteen should be used in preference to the coarser mohair or Utrecht velvet. The covering of the back (inside of course) will present no difficulty ; but it will not be easy to cover the whole of the shelves and ends without leaving the front edges in an apparently unfinished state, from the liability of the velvet to fray out. To avoid this, therefore, let the velvet only approach to within 2in. or 3in. of the front of the shelves, and cover the ends in the same line, without attempting to cut the material so that the shaped curves. The velvet will still be apt to fray out, so partly to prevent this, and partly to give a finish to the work, nail small half round beads over the edges. It may be a "wrinkle" to some

FIG. 143.—ANOTHER ALTERNATIVE.

that these beads may very easily be formed by splitting ordinary cane in half. The knots, or rather extra thickness, may be reduced by scraping. The cane may be painted as it is, but the paint is less liable to chip off if the surface be rubbed down a little beforehand. The pieces can easily be fastened down with small wire nails or brads. By arranging that they stop just under the curved pieces, or spandrels, a "patchy" effect may easily be avoided. Fig. 144 shows one of the recesses so lined and finished. Instead of painting the beadings, whether of cane or slips of wood rounded off, they may be covered with velvet. Glue the pieces well over and then put the velvet on so that the joint is underneath. By driving the nails through the cane before covering it, the heads of these will not be visible, a few taps with the hammer sufficing to fix the covered moulding down. In fixing plush to a glued surface do not press it down too much, or handle it more than necessary till the glue is dry, otherwise this may come through and injure the surface. Press it down quickly and firmly and no harm will result. If the joints of the velvet at the other angles are not neat they may be covered with similar

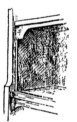

FIG. 144.—RECESS, LINED AND BEADED.

strips of cane, either covered with paint or velvet, but quartered instead of halved.

The top edges of the ends may present a difficulty. The way to manage will be simply to carry the velvet on to the top edge, say to about the middle of their thickness, and to hide the joint by another of the covered slips of beading, which it will be just as well to trim off neatly in front and mitre at the back corners.

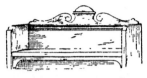

FIG. 145.—ORNAMENTAL TOP.

Looking-glasses may easily be fixed behind each of the shelves, except the top one, by means of quartered cane Let the glasses be cut the exact width, so that they fit in accurately between the ends. The fit the other way is not so important, as the overhanging shelves will prevent the upper edges being seen, and the lower ones will of course rest on the shelves below. To prevent the glasses falling forward, or in other words to fix them, it is only necessary to use three strips of cane to each : one along the bottom edge and the other two upright.

By the use of small, neat mouldings, fixed along the fronts of the shelves with glue, a good finish may be made to these parts by those who do not like the look of square edges. The top of the back may also be ornamented, or at any rate its stiffness relieved by shaping it, as shown in Fig. 145. By such simple means as these the plain, easily-made whatnot may be made, not only a useful, but a really handsome piece of furniture, without to any appreciable extent increasing the difficulties of construction.

AN UMBRELLA-STAND.

THE umbrella and stick stand which forms the subject of this chapter is very easily made, and as an article of hall furniture will be found very useful. It will accommodate eight umbrellas and eight sticks, whips, &c., the vase (C, Fig. 146) will also be found useful for holding cards, gloves, &c.

The central pedestal or shaft (A, Fig. 146) should be 2ft. long and 3in. in diameter at its thickest parts; it may be turned to any pattern according to the fancy of the maker, but care will be necessary to avoid weakening the ends, as these must be sufficiently strong to allow of the insertion of a ¾in. or 1in.

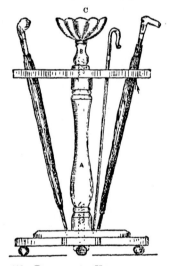

FIG. 146.—DESIGN FOR UMBRELLA-STAND.

wooden screw to a depth of 4in. or 5in. at each end. Any of the hard woods usually employed for such articles of furniture will suit for this stand, but oak should be procured if possible; if for any reason oak cannot be obtained, well-seasoned beech can be used.

Should the worker be able to procure an old-fashioned bed-post, he might use it with great advantage for the pedestal, as some of these are beautifully turned and often carved.

Having made the shaft, the next step will be to make the two wooden screws.

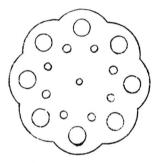

FIG. 147.—PERFORATED LEAF OF STAND.

For these a screw-box of a suitable size will be required. Beech will be best for the screws. Having prepared two pieces about 12in. long and a little more than 1in. in diameter, turn them up to the exact size for the screw-box, and, say, 9in. long. Now pass the ends through the box, making a screw 3½in. long on each end. Next bore the ends of the shaft with a suitable bit—*i.e.*, about 1/16in. smaller than the diameter of the screw—and tap the holes to a depth of 3½in.; insert a screw in each end to this depth, having first applied a little strong glue, as the screws are to be fixtures in the shaft.

The next item will be the small pedestal (B, Fig. 146), which should be about 6in. long, and also sufficiently strong to bear boring and tapping, as in the case of A. This is to act as a nut by which to secure the perforated top (Fig. 147), and is a support also for the vase (C, Fig. 146). The lower end is treated exactly as in the case of the ends of A, and the upper end should be turned slightly concave to receive the bottom of the vase, which is to be attached to it eventually.

The base of the stand (Fig. 148) comes next. It consists of a disc of 1¼in. stuff, chamfered at the edge, to the bottom of which the three pieces A, B, and C (Fig. 148) are fixed in the position indicated. These pieces are also of 1¼in. stuff, and are attached to the disc from underneath with glue and stout 2in. screws. The projecting ends of these foot-pieces are nearly semicircular; they are about 6in. wide, and sufficiently long to project about 4in. or 5in. beyond the edge of the circular piece.

Each foot-piece is provided with a round knob, about 1½in. in diameter, turned with a shank 1in. long; the knobs are fixed with glue in the positions indicated at Fig. 146. Casters may of course be substituted for these latter should the maker prefer them, but he

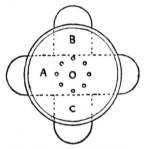

FIG. 148.—BASE OF STAND.

can make the knobs for nothing, and they will be more in keeping with this stand. Now bore a hole in the centre of the base, through which the wooden screw of the pedestal (A, Fig. 146) is to pass. This hole may be tapped, but perhaps a better method will be to bore it large enough to permit the screw to pass through—making a good fit, but no thread—and to secure the screw underneath by means of a square nut of beech, about 2in. by 1¼in.

We now come to the making of the perforated top (Fig. 147), which is formed from a disc 24in. in diameter, of 1¼in. stuff—finished work—pierced by eight holes 3in. in diameter, and placed 1½in. from the circumference and equidistant from one another; also by eight smaller

holes 1½in. in diameter, placed about 6in. from the circumference, and in the positions indicated—*i.e.*, alternating with the larger holes.

The edge of this top may be left circular and finished with a simple bead, made by a scratch-router, or reeding-tool, or it may be shaped as suggested at Fig. 147, or otherwise, according to taste.

The holes to receive umbrellas, sticks, &c., should be carefully cut, the edges of the larger ones being slightly rounded with a fine wood file and glass-paper, making them as smooth as possible, otherwise they would fray the silk of umbrellas. The smaller holes may be cut clean and sharp to the exact size if the worker is fortunate enough to possess one of Clarke's expansive bits. Indeed, with a larger size of this tool, the 3in. holes might also be bored.

Now bore a hole in the centre of the top for the wooden screw upon the upper end of the pedestal (A, Fig. 146) to pass through; screw on the small pedestal (B) until it grips tightly, and we are ready to consider the last item—viz., the vase (C). This may, of course, be turned by the maker himself, but a good deal of trouble and time may be saved by utilising a common wooden dish or basin, as used for butter-making and other domestic purposes. These can easily be procured for a few pence, and may be ornamented sufficiently for our purpose—a dish about 8in. or 9in. in diameter will suit best; it may be fluted and mitred on the rim, as suggested in the cut, or otherwise carved, &c. A stout, round-headed screw, about 2in. long, with a small washer under the head, and a little glue, will be found efficient for fixing the vase upon the pedestal (B, Fig. 146).

This stand will be improved by having two semi-circular dishes of stout tin made to stand upon the circular disc of the base and close round the pedestal, to receive the drips from umbrellas. This can easily be procured from the nearest tinman should the maker not be skilled in the craft himself.

The work should be kept perfectly clean and free from finger-marks, grease, &c., as whatever treatment is employed with regard to polishing, painting, &c., cleanliness is essential.

If made of good oak the stand may be merely oiled or polished, or better still, if the worker is "up to it," it might be "smoked" in ammonia fumes and French-polished. Ebonising and polishing will also suit, should the stand be of beech.

A LUGGAGE-STOOL.

THOSE who have occasion to fre-quently use a portmanteau, or the more modern Gladstone bag, know very well the awkwardness of trying to pack or unpack it with a chair as support, and it is not a little remarkable that the

FIG. 149.—LUGGAGE-STOOL.

luggage-stool is so rarely regarded as a part of the ordinary domestic furniture. It is neither costly nor cumbersome, while, when covered with a movable cushion, it may form a most convenient and sightly small ottoman.

A convenient size to make the luggage-stool (Fig. 149) will be about 2ft. 6in. long by 1ft. 6in. wide and 1ft. 6in. high, or ordinary chair height. These dimensions can of course be altered to suit special wants.

The top (Fig. 150) may receive the first attention. It consists of a framing with cross-rails. The former is com-posed of 1¼in. stuff, and the latter may be of the same, or a little thinner. In width the pieces are 2½in. This is the full limit necessary, and the frame should not be formed from anything narrower, though the rails may be less. In fact, if of sufficient thick-ness, they may be "anything," though it is not advisable to have them too narrow, as the amount of work is proportionately increased. With laths of the width named it is also easier to make the speediest form of joint, viz., that with dowels.

The dowelled joint is sometimes objected to on the ground that it is not permissible in good work. Of course, a good deal depends upon the way the joint is constructed, for there can be no doubt that a badly-fitted dowel is weak and objectionable, but the same may be said of any method of construction, however good it may be when properly done. In itself, and regarded without prejudice, there is nothing to be urged against dowelling on the score of either dura-bility or strength; while with a very ordinary amount of skill the joint can be made very quickly.

Before the joint can be made, the parts necessary for the top must be got out. Two pieces for sides will be wanted: these must be of equal length, with ends properly squared off. All the pieces of the top must be truly square at both ends. Unless they are, it will be futile to expect that good close-fitting joints can be found. A shooting-board will greatly facilitate accurate working-up; but, failing this, free use of the square must be made in testing the work as it proceeds.

The side pieces of the frame will be 2ft. 6in. long. In addition to them two other pieces for the ends, which should be of the same dimensions in everything except length, will also be required. As the width of the top is 1ft. 6in., we have merely to deduct the double width of the two sides, or, say, 5in., to arrive at the length of the ends, which will therefore be 1ft. 1in. long. This will also be the length for the laths, or cross rails, and the necessity for these being exactly of equal length cannot be too strongly insisted on. Reckoning in the laths, seven pieces altogether form the top.

Now for a dowelled joint. The dowel is a pin of wood inserted into holes bored into both the surfaces to be connected. In order that the dowel shall hold firmly, glue is used along

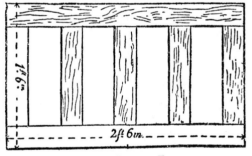

FIG. 150.—PLAN OF TOP.

with it. Fig. 151 shows one end of a cross-rail, with the dowel inserted in it. The dowels are, in reality, nothing but round tenons, and instead of a mortice being cut in the piece to be joined on, holes are bored with a bit. It is, of course, essential that the bit used must make a hole of exactly the same diameter as the dowel-pin. The holes must be perfectly perpendicular, so that to get a perfect joint there is no strain on the dowels, as there is bound to be if two holes opposite to each other happen to run at a slight angle from each other. The dowel pin for general cabinet work is about ⅜in.; but the

worker can make his dowels to suit any convenient bit he may have. The usual way to make the dowels, or rather, to shape the wood from which

FIG. 151.—END OF RAIL WITH DOWELS.

the pins are cut as required, is, after roughly shaping sticks of a convenient length, to hammer them through a hole of exactly the required size in an iron plate. The wood should be of some hard, tough kind, and nothing is better than beech for general purposes. It must be thoroughly dried before it is used, in order that the pins may not shrink, and so become loose after the joints are formed.

In the work in hand, first of all the exact position of the holes must be marked in the ends of the laths and in the pieces to be attached. Set the gauge to about half the thickness of the stuff, and with it mark across each end of the pieces, working from the face side of the wood, or, at any rate, from the same side of all, and to prevent mistakes it will be just as well to mark this with pencil, or in some other way. Those who have a proper bench may now fix up the pieces together in the bench-screw, with ends upwards and all perfectly level; then with the gauge-block against the edges mark across the ends. Repeat from the other edges. There is then on the end of each piece three lines, the point of intersection giving the centres for the holes. These may be about ¾in. deep, and of course are bored while the pieces are still in the bench-screw. When the holes have been made in one end the same routine is gone through with the other ends. With the gauge still set to mark the same space, the inner edges of the top must be marked from end to end. With the aid of the square or compass, then set out on these lines the points for centres of holes, and bore them, using of course the same bit as before. The glue being hot and ready and the dowel-sticks handy, bruise an end of one of these by knocking it slightly at the edges, after having, if the sticks are perfectly round, taken a shaving or two off the whole length. The object of

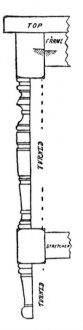

FIG. 152.—DIAGRAM OF LEG AND FITTING.

doing this is to afford a means of escape for the air which would otherwise be compressed when the dowel is being driven home. The hammering of the end is to round this part slightly and give it easy entrance. Next glue the inside of the hole, and without loss of time, hammer the dowel home. Some glue will exude, and it should not be allowed to harden, though if the work proceeds rapidly it will have no time to become too hard to prevent the wood from being closely jointed.

With the saw cut the stick off, leaving about as much projecting, rather less than more, as the depth of the corresponding hole in the side piece.

Having done this, a dowel is to be inserted in a similar manner into each hole of the laths. When this has been done the top may be put together by gluing the holes in the sides and fastening the laths and ends to them. If a couple of cramps are available they will be useful in making close joints, but if they cannot be used, the laths, &c., must be knocked home, taking care in doing so not to bruise the edges of any of the pieces. If it is not found convenient to round the projecting ends of the dowels by hammering them before attaching the sides, a file may be used instead, though it is hardly so good. The dowels should, as nearly as possible, completely fill the holes, and the hammered ends are better than the filed, inasmuch as under the influence of the moisture of the glue they expand to their original condition. After the glue has hardened, the sharp outer edges on the top of the frame should be rounded off a little, or, if preferred, they may be bevelled.

Attention may next be given to the legs and stretchers. They are formed from 2in. squares, which will be, when finished, about 1¾in. Figs. 152, 153, and 154 give the outlines on a scale, which can easily be enlarged to full size. The turning is of a very simple character, and can be executed with the few ordinary tools which anyone having a lathe is sure to possess. The parts should not be cut

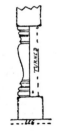

FIG. 153.—PATTERN OF END STRETCHERS.

to the exact length before turning, as it will be found much better for various reasons to trim the ends off afterwards.

The three stretchers may be of lighter stuff than the legs, but care should be taken that the squares shoulder up well to each other. The union may be effected by means of dowels, or by mortice and tenon.

The parts represented as turned might be squared and chamfered instead, the legs being tapered to the bottom, as shown in Fig. 155. Many varieties of this simple means of finishing off will doubtless occur to the worker who cannot manage turning.

To prevent the stand from looking too heavy the stretchers themselves, if unturned, may very well be formed of 1in. or even ¾in. stuff, about 1½in. in width. The long middle stretcher may be simply screwed on to those at the end, or attached in the way already stated.

The underframing or stretchers at the top of the legs may be 2in. wide and 1in.

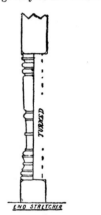

Fig. 154.—Portion of Long Stretcher Connecting Ends.

thick, though less will do. The parts should be set back within the legs to the extent of ⅛in., instead of being flush, the legs themselves being overhung by the top. As the framing of this is 2½in. wide and the legs will not be more than 1¾in., this can easily be arranged for, the result being much more gratifying than if all the parts finished on the same line.

To attach the top to the lower part of the stand, long screws may be driven through the framing underneath. By making the holes for a certain depth sufficiently wide to admit the heads of the screws, as shown by the section (Fig. 156), no very long screw will be required. Those who prefer to do so, may

put pockets for the screws, and drive them in slantingly from the inner side of the framing.

Luggage-stools are often put together in a somewhat different manner, the framing at the top of the legs being dispensed with. This is a simple construc-

Fig. 155.—Alternative Leg, &c.

tion, but the appearance is hardly so good. The legs simply have to be turned with a pin on their upper ends, to fit into holes made to receive them in the framing.

The wood to be used should be chosen with some regard to the remainder of the furniture in the room where the stool is to be when finished. It is very common to find luggage-stools made of beech and stained to imitate

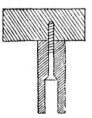

Fig. 156.—Diagram of Screw Sunk in Frame.

walnut, mahogany, &c. If this is carefully done the appearance is very good. French polish is much to be preferred to varnish, which, do what one will,

H

always has a slovenly look on good furniture. Polishing takes more time and is more difficult than varnishing, but the result is so far superior that it seems a pity that it is not more generally adopted by amateurs.

A MORESQUE MUSIC-HOLDER OR BOOKSTAND.

THIS piece of furniture is a very useful addition to a drawing-room. But the shape, size, and construction may be modified to form other and equally effective articles, such as newspaper racks, portfolio-holders, bookstands, &c., as suggested in the sketches, Figs. 157 and 158.

The following directions are for the construction of a folding rack or music-holder out of one solid piece of timber, which is one of the nicest pieces of

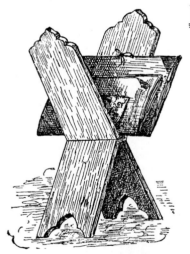

FIG. 157.—SUGGESTIVE SKETCH OF MORES-QUE MUSIC OR PORTFOLIO HOLDER.

carpentry one can well try; not that it is difficult to anyone possessed of

ordinary skill in this particular line, but few know how it is accomplished.

The drawings, Figs. 159, 160, and 161, are not drawn to scale, but are slightly

FIG. 158.—SUGGESTIVE SKETCH OF MORESQUE BOOKSTAND.

exaggerated in detail in order to enable the reader to see at a glance how the thing is done ; while, therefore, we refer to the figures as we proceed, we shall do so only with reference to shape. The measurements mentioned are those for a music-holder 30in. high when folded, and 15in. wide.

Procure a piece of well-seasoned beech, 30in. long, 15in. wide, and 1½in. thick; select wood which is perfectly free from knots and flaws, and plane it up quite square and true every way. Now measure 12in. from one end, and with a square draw the line L K (Fig. 159), and continue it round the four sides. Next draw a second line 1in. from L K, and continue it also.

With a gauge mark the lines C A, A B, B D, and E G, G H, H F. Now with a fine and very sharp hand-saw make a cut as indicated by the heavy dotted lines A B C D, beginning at A B and cutting down to C D. Great care will be required during this operation; especially must the bottom of the cut,

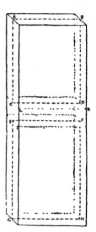

FIG. 159.—PIECE OF TIMBER SHOWING SAW CUTS.

C D, be perfectly straight, and not lower than the position indicated.

The wood must be held very firmly by a strong bench-screw or vice. Now reverse and cut E G H F.

The wood is now divided evenly into two boards, except that portion lying between C and E, D and F, which is 1½in. by 1in. by 15in., and is indicated by the light dotted lines (Fig. 159). This portion is to form the hinge, as represented at Fig. 161, which, as stated above, is a sketch in perspective, not drawn to scale, and meant only to give an idea of the construction of the hinge.

FIG. 160.—METHOD OF CUTTING HINGE.

Fig. 160 shows the method of cutting the hinge. The lines I J and L K, already drawn upon our work in hand, and the corresponding lines upon the

other side, are now divided into five equal parts (of course there may be any odd number of divisions in the hinge), and the lines *a e*, *b f*, *c g*, and *d h* drawn.

Next, these lines must be sawn through from side to side. At the points *a*, *b*, *c*, and *d*, bore small holes with a suitable bradawl, and with an ordinary fretsaw to make the opening, and then a fine keyhole-saw, cut out the lines perfectly straight, as marked on both sides.

Now the line L *a* is to be cut down, at right angles to the surface, until it meets the line C D (Fig. 159), *i.e.*, when it has cut through half the thickness of the wood. This should be done with a very sharp chisel, and at the same

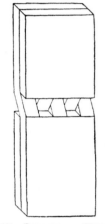

FIG. 161.—DETAILS OF HINGE IN PERSPECTIVE.

time the portion L *a e* I is sloped gradually from the line I *e* (see corresponding part in Fig. 161). Now cut the line *e f* (Fig. 160) through until it meets E F (Fig. 159), or half the thickness of the wood, gradually sloping *a b e f* from the line *a b*. Treat the remaining portions alternately in the same way, and when this is accomplished turn the work.

Assuming that the workman has his work similarly marked upon the reverse side, the same lettering, &c. (as in Fig. 160), will still be available if the work be not only turned, but also the position so changed that the end which before was

H 2

toward the worker is now away from him, and *vice versâ ;* in other words, turn the work longitudinally and not laterally. The corresponding line to I *e* will now be in the position L *a*, that corresponding to *e f* will agree with *a b*, and so on; therefore the same directions already given for cutting the hinges on the one side may now be followed for the other, using exactly the same lettering.

As these operations are carried out the worker will find that the solid board is being transformed into two leaves, inseparably joined by a hinge having five divisions; and with the aid of a very sharp knife, or, better still, a Swiss knife, as used in wood-carving, he will be enabled to free the hinges completely. If he be one who has never happened to see this "feat" in carpentry performed before, he will, we venture to predict, be filled with the most lively satisfaction.

The extent to which the rack will open can be regulated by the width of the space marked off in the first instance for the hinges; the greater the space, the wider will the finished article open, but for a music-holder, &c., the best angle is about 125deg., *i.e.*, the outer angle made by the sides.

The thickness of the stuff will also, of course, regulate the width of the hinges, but a safe width will be about two-thirds of the thickness of the timber when squared up at the time of marking.

The work should now be planed upon the inside, as far as the plane can go, the inaccessible parts being finished with glass- and sand-paper.

With regard to the after-treatment, whether the maker will shape the ends as suggested at Fig. 157 or leave them square, or otherwise ornament his work, we leave to his own decision. Great things can now be done in this direction with the aid of Aspinall's Enamel, Japanese paper, &c.

Fig. 158 is a suggestion for a bookstand upon this principle, and if made of rather stout mahogany, and French polished or carved, makes a very useful and ornamental adjunct to a library table, &c.

A COMBINATION MUSIC-CABINET AND COAL-BOX.

THIS curious piece of combination furniture—coal-box and music-cabinet—is introduced to amateurs as a novelty. The plan on which the former is made is not new, though the manner of fixing the coal-receiver and the front is not common, and simpler than that of hanging the coal-box to the sides by means of swivels or swing-joints. This particular piece of furniture is very easy to make, and yet, if carefully put together and nicely finished, is a very creditable article for either dining- or drawing-room.

Fig. 162 shows a general view of the piece of furniture. The measurements are as follow : Bottom board, 18in. long by 14in. deep, of ⅝in. or ¾in. wood ; sides of box, 2ft. high, 12in. wide, ¾in. thick ; the top, 21in. long, 13in. wide, ¾in. thick. The wood used may be one of the following : oak, mahogany, walnut, pine, or ordinary deal.

The sides are fixed to the bottom by screws from underneath, and should have a clear space of 13½in. between them, also leaving a margin of 2in. of the bottom board in front; this will bring the sides flush with the bottom board at the back. The top board is screwed on to the sides in like manner, overlapping them in front 1in. and each side about 3in. Four fancy brackets, 4in. by 3in., as shown supporting the top board, are more for finish, so need not be fixed very firmly, ½in. wood being used for these. The front of the box must be of substantial thickness—certainly not less than ¾in.—as it will have to stand the weight of the coal-box, and should be "made up"—by which is meant, not cut out of a single board, but having two uprights and two cross-rails. The uprights are 4in. wide, the cross-rails 6¼in. ; this leaves an open space measuring 11½in. by 5½in. to be filled in according to fancy. A good plan is to get a hand-painted tile, 12in. by 6in., or two 6in. fancy tiles of some delicate tints, letting them into a rebate

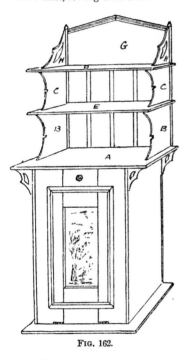

FIG. 162.

at the back ½in. deep, the space admitting of a ¼in. margin on all sides. The uprights and cross-rails are mortised and tenoned.

In making the front, it is as well to let it be full large, so that when fixed together it can more easily be squared up and fitted into its position, two thin fillets being screwed to the sides for the

front to shut against. Two strong brass hinges, 2½in. long, are next let into the bottom edge of front, so that they

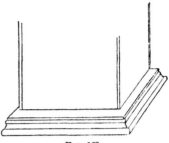

FIG. 163.

are flush with the surface when shut together (the knuckle edge being outwards) ; they are then screwed on to the bottom board—and this must be done before the back is fixed on. The width of the uprights and cross-rails, if left plain, would sink the centre panel into insignificance, so a fancy moulding 1in. wide, mitred at the corners, and fixed on about 2in. from the sides and 3in. from the top and bottom, greatly relieves the flatness, the moulding being fixed with thin glue and needle-points. So much for the coal-box case.

Should the amateur consider the whole piece of furniture rather dwarf in appearance, as it will not be higher than about 3ft. 9in., he can add a false base.

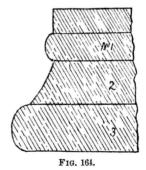

FIG. 164.

Fig. 163 shows a plan for a false base, which can be made up as follows, after the sides have been fixed to the bottom board.

Three pieces of wood, 4in. wide, of sufficient length to form the front and two sides, are required, and as the corners are mitred this must be allowed

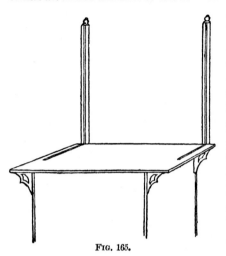

FIG. 165.

for when cutting out the wood. Fig. 164 gives a section showing how the base is made up. First a piece ¾in. thick (No. 1), with the front edge rounded, is screwed to the bottom board with mitred corners. No. 2 is 1in. thick, hollowed on the front edge, and screwed to No. 1. No. 3 is also 1in. thick, rounded on the front edge, and screwed to No. 2; all screw holes are well countersunk, so that each board fits flat against the other, so as to resemble one block. By referring to Fig. 163 it will be seen the base projects slightly beyond the bottom board. Fig. 165 shows the uprights forming the sole supports for the shelves and side-pieces of the top part for the music cabinet. The uprights are 1ft. 9in. long, of ¾in. stuff 1¼in. wide, the front edges being beaded and a groove $\frac{5}{16}$in. wide, ¼in. or ⅜in. deep, cut down the centre. This is done by the aid of the "plough"; but in the event of the amateur not possessing this particular plane, a simple method is to get a marking-gauge and set it so that when marking from both edges of the wood it will leave a space

$\frac{5}{16}$in. wide in the centre; then lengthen the marker and, by filing it flat like the head of a spear, it will form a cutter sufficiently good to cut in the required depth. Then a ¼in. chisel will take out the centre. The uprights should then be screwed to the back of the top board, 1in. from either end; but before fixing them, grooves should be cut in the top board, to correspond with those on the uprights and in a line with the same for the side-pieces (B) to fit into, and should be 9in. in length. These grooves can more readily be cut out with a chisel, as they are "cross-grain." The shelves and side-pieces should then be cut out; shelf E, the lower one, being 21in. long, 10in. wide, ¾in. thick; shelf D, the top one, being 1in. less in width. These must be grooved on both sides and at both ends, for the side-pieces to fit into. The lower ones (B) are 9in. back to front, not including letting in, which must be allowed for when cutting them out; and those marked C are 8in., so that the grooves on the shelves must be cut accordingly.

To fix the shelves and side-pieces, the latter are glued into the grooves, and the former in addition should be strengthened by screws through the uprights at the back. I should mention that the top grooves on shelf D, into which the side-

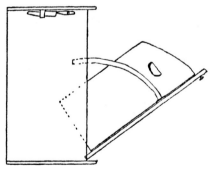

FIG. 166.

pieces marked H are fixed, as well as into the uprights, are only 7in. long, and they should taper to the top of the uprights, the pattern of them being in character with the small brackets under the top board A. The side-pieces are ⅜in. thick, the lower ones being 8in. high when fitted, and those marked C, 7in. The height of pieces H will depend upon the space left after the others are fixed. The back (G) only extends down as far as the top shelf, and can be of thin wood (say ⅜in.), which is screwed on to the uprights at the back, the top corners being flush with the uprights; the point in the centre should be about 3in. higher. A moulding or beading ¾in. wide, ¼in. thick, forms a finish for the top of the back, and is fixed on with screws from behind. The knobs on the top of the uprights can be of wood or metal. Should the amateur not possess a lathe, he can get some very effective brass knobs, such as are fixed at the ends of picture-rods, to form a finish.

All edges showing, which are drawn as square, should have a beading on them, which is easily done by means of the "bead-router." The beading helps to relieve the appearance where the boards are obliged to be thick—such as the shelves, for instance. To prevent the music from slipping through at the back, three laths, 1in. wide and ⅛in. thick, should be screwed to the shelves and top board at back; they also help to strengthen the stand.

Fig. 166 shows a section of the coal-box and position of the coal-receiver when pulled out for use. The receiver should be of the following size, strictly outside measurements: Height at front, 18in.; ditto at back, 15in.; width, 11¼in.; back

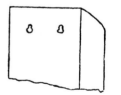

FIG. 167.

to front, 10in. This of course can be made of wood, but it is better to get one made in galvanised iron, as there is no fear of warping through wet coal, and an iron one

will last a dozen wooden ones. This is hung on to the door, and the receiver should have two holes cut in the front (as shown in Fig. 167), the same shape as

FIG. 168.

mirror-plates, so that it can be easily taken out if required for filling. Two stout, round-headed screws, 1in. long, are fixed in the door as far as they will go without the points coming through, and the receiver hung on these. To prevent the door from falling out, and to hold it in a proper position when open for use, what is known as a quadrant-stay should be fixed on both sides. Fig. 168 will give the amateur an idea as to what is meant by a quadrant-stay, which he will probably have to get made in iron or brass. The plate A is a fixture on the stay, and is screwed to the door; the plate B is a guide for the stay, and is screwed to the side of the coal-box. In case of two being used they will have to be made handed, i.e., one right and one left hand, the end being turned up slightly to form a stop, and this coming against plate B will prevent the receiver from coming out too far. On each side of the receiver a drop handle of iron wire should be fixed for the convenience of lifting. A small piece will have to be cut out of the fillets on the side to allow for the stay moving in and out, and there should be a space of 1in. on either side between the receiver and side of coal-box to allow for these stays. A strap of brass or hoop-iron, screwed inside to the top board, will hold the coal shovel, and this will complete this handy piece of furniture, with the exception of a small brass knob to open the door with. The weight of the receiver and its contents will be sufficient to keep the door closed. However, a small spring catch could be fixed on the top edge if thought necessary. The question of finish must be left to individual taste.

A COMBINATION MUSIC-BOX AND STOOL.

THIS is not a new idea, but it is one which should commend itself to such as are anxious to make as much as they can for themselves, as the work is quite within the scope of an amateur. The only parts likely to present a difficulty in making are the legs, but if the amateur is the possessor of a lathe the difficulty is soon overcome ; if he is not, it will be quite excusable for him to get them done either by a friend or by a professional turner.

In the article under consideration, the top of the stool forms the lid of the box, and it will be noticed by referring to Fig. 170—which shows the stool minus the legs—that the top, or lid, slopes upwards towards the back, where it is hinged to the box. This idea will commend itself to all pianoforte players, the incline position being in itself the more comfortable, besides giving a greater command over the instrument. In the

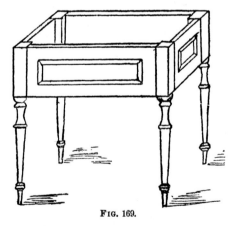

FIG. 169.

most modern type of music-stools, which are being made now of light ironwork, artistically bronzed or gilded, the inclined seat is much used.

Commencing with Fig. 169. The legs are 1ft. 9in. high, turned out of wood 1½in. square. Fig. 171 shows a leg more clearly, though the pattern can be altered

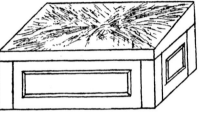

FIG. 170.

to suit the taste. The top, 6in., into which the sides of the box are mortised, is much improved by the flutes as shown, which should be cut in the two outer sides of the square, and are easily done by means of the bead-router. The box when made measures 20in. long, 14in. wide, extreme measurements, and 6in. deep. This is of course without the lid. The front and back will therefore measure 17in. long between the tenons, and the sides 11in. between the tenons. All four sides must be fitted so as to come flush with the outsides of the legs. Fig. 172 shows a cross-section of a leg with the sides mortised into it ; it also shows the position of the tenons, which, by the way, must not appear at the top of leg, but should be cut ⅓in. from top and bottom.

Walnut wood should be used in making this stool, American walnut being comparatively cheap, though rather lighter in colour than English. The material for the legs can be obtained cut the proper size, viz., 1½in. square ; the sides can be of any thickness, from ½in. to ¾in., the thicker the better ; the tenons should be about ¾in. long, well fitted and glued in, and fixed with screws from the inside. The bottom of the box may be of deal, ½in. thick, and must not be visible below the sides, but should be laid on fillets fixed on the inside of sides.

The top, or lid, should be made in the form of a frame, the wood not to be less than ¾in. thick, the front of frame being 1in. deep, and the back 3in. (Fig. 173). The corners should be mitred and glued together, and further strengthened with triangular pieces, screwed to the frame from the inside. Round the bottom of the frame a ½in. or ¾in. bead-moulding will greatly add to the finish.

For covering the frame for the seat, the foundation is chair-webbing inter-

can be easily obtained at upholsterers'. Inside the lid, to hide the webbing, some delicate tinted material will make a good finish. The lid is hinged to the box with two 2½in. brass hinges, and a tape stop should be fixed inside, one end to the

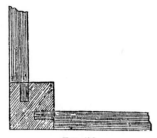

FIG. 172.

box, the other end to the lid, so that when the latter is open it will prevent its going back too far, which would strain the hinges. On the outside of the box on all four sides, if the amateur thinks the appearance too plain, panels of thin moulding will have a good effect.

The music-box and stool can either be varnished or French polished ; in either case it should be done before the seating. Should the amateur find, when the box

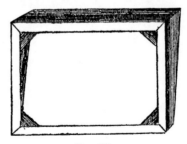

FIG. 173.

FIG. 171.

laced and nailed to the top edge of the frame, and stretched as tightly as possible, and over this some canvas wrappering, such as is used in seating chairs ; then the padding, which should be of horsehair, flock, wool, or any like material. The outer covering must of course be left more to the discretion or taste of the maker—leather, plush, or what is known as " saddle-bag " being equally suitable —and finally a gimp edging to hide any imperfections of finish. All these things

is filled with music, that it is rather heavy to move about, casters on the two front legs only will help it very much. These will necessitate the legs on which they are fixed being slightly shortened according to the height of the casters.

A CORNER-BRACKET CABINET.

THE woodwork of this bracket, &c. (Fig. 174), may be of the kind that comes handiest if it be fairly workable and sound; but beech, as it works beautifully and is capable of a very high finish, is recommended.

Fig. 175 shows the shape and proportions of the sides of the bracket. They are $\frac{1}{2}$in. thick, finished work, and one side is $\frac{1}{2}$in. wider than the other, the vertical dotted line representing the difference in size. The curved or "shaped" portions of each side should be chamfered on the inside, and the straight

FIG. 174.

FIG. 175.

portions in the centre should be left rather "full" when cutting out, in order to provide for bevelling, so that the front edges shall be flush with the door, &c.

Fig. 176 gives shape and proportions of the two larger shelves, which form the top and bottom of the cupboard; they are of the same thickness as the sides.

Having cut out and carefully planed the sides and shelves, finishing with fine glass-paper, these parts may be put together. This may be done by nailing or screwing the wider side to the other, the back edge of the narrow side being placed against the space indicated by

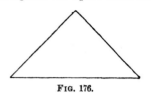

FIG. 176.

the upright dotted line. If nails are used, they should be 1½in. wrought brads or "battens," but in case beech is the timber employed, rather thin 1¼in. screws, the heads being countersunk, will answer better than nails.

The shelves should now be secured in position, as shown by the transverse dotted lines in Fig. 175. The same remarks just made apply here also.

The next items are the narrow panels at each side of the door. First prepare two slips 12in. long, or according to size of the bracket, by 1½in. wide. Place these 3in. from the side, and at right angles to the front edge of the shelves, their outer edge being flush with the same. Secure them by two battens at each end driven through the shelves, being very careful to avoid splitting.

Next prepare two narrow panels to fill the spaces made by these. These panels should fit the openings exactly; they may be of pine or any light wood, and are secured by very slender brads;

FIG. 177.

place them so that there shall be a margin of about ¼in. between them and the edge of the shelves, &c.

The door consists of one piece, cut to fit the opening rather loosely. The

right hand edge is rounded to admit of the door turning on its hinge, which simply consists of a piece of iron or brass wire (No. 8, B.W.G. or about $\frac{3}{32}$in. thick), about 2in. long, passing through the shelves, and inserted in the door at top and bottom. Fig. 177 shows plan of part of the upper and lower edges of the door, with the position of the holes which receive these wires.

If these holes are carefully bored, this makes a good hinge; and if a small plate of sheet brass (No. 22, B.W.G.) be fitted to the door, with a hole for the wire, the hinge will be greatly improved; but if hard wood is used this will not be necessary for such very small doors.

Now glue a small piece of wood about ⅛in. thick upon the side of the other

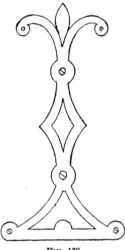

FIG. 178.

upright piece, to act as a stop for the door, so that when shut it shall correspond with the panels at each side, i.e., leaving a margin of ¼in. formed by shelves and two upright pieces.

The small shelf below the cupboard hardly requires any directions. Its sides are about 6in. long, and the front may be either straight or forming an arc of a circle, or otherwise shaped according to fancy. It is secured to the bracket in the same way as the other shelves.

The narrow panels may be fitted with

repoussé brass or with strips of Lincrusta Walton, as suggested in the sketch, Fig. 174.

The bracket is now made, as far as the woodwork is concerned, and if the worker has taken pains and fitted the various parts together accurately, the result should be satisfactory, and we are ready to proceed with the mountings.

Space compels us to condense these directions as much as possible.

Fig. 178 is a simple design for an ornamental hinge, and should be drawn the required size on paper, and then transferred to a piece of sheet brass—No. 23 or 24 B.W.G.

The brass should first be thoroughly cleaned and polished with very fine emery cloth—Oakey's No.0—on the side upon which the design is to be drawn; the other side should be well rubbed with coarser emery cloth until the whole surface becomes quite bright. This is necessary to prepare it for receiving the solder, to be presently described.

Having transferred the design to the brass, which should be sufficiently large to allow a margin of about ½in. all round the design, with a fine bradawl and hammer punch holes all round the margin about 1in. apart. Now tack the metal upon a rather thick, smooth board, or better, upon the surface of a smooth block of hard wood, using ⅜in. "cut" tacks. Next, with a ¼in. "tracer," as used in repoussé, indent all the lines of the design, at first rather lightly, and then heavily; a curved "tracer" will be required for the curved portions. When the whole design has been thus treated, take another ¼in. tracer, which has a sharp edge, and go over the outline until the brass is cut through. Place the hinge upon a flat bench, &c., and by gently tapping with the hammer remove any tendency to "buckle" or twist, and if all has gone well the hinge "in the rough," but clearly cut, and rounded on the upper side, will be completed.

The next operation is to run solder (common plumber's) into the back, or concave side, of the hinge.

First coat the brass with "killed spirit" (spirits of salts, *i.e.*, hydrochloric acid in which zinc is dissolved until it will dissolve no more). Heat a soldering iron, and holding a stick of solder in the left hand and applying the iron, let the solder run in until the whole of the hinge is filled. Take care not to allow the melted solder to overflow, or it may adhere to the right side of the work. This may be guarded against at first by coating the right side with black lead and beer. When this operation is concluded, the rough edge of the hinge is filed smooth and the design brought out clear and sharp, and it is now ready to be "chased."

Here there is unlimited scope for artistic work, and many and varied effects may be produced; but for the present example we suggest a rough frosted surface. This is effected by the use of a rather coarse "matting" tool, consisting of a punch having several points upon the face. Place the hinge upon a flat piece of thick lead or upon a solid block of hard wood, and, holding the "mat" in the left and the hammer in the right hand, proceed to roughen the whole surface by allowing the indentations to overlap each other, and by moving the tool hither and thither over the work. If the points upon the face of the matting-tool are tolerably sharp—a star-shaped face is best—the result will be a rather rough but bright frosted surface.

Repeat the above operations for a second hinge; drill holes as suggested in the sketch for small round-headed brass screws, by which the hinges are attached to the door, and this part of the work is done.

Fig. 179 is a simple design for a key-hole escutcheon, which is to be treated in the same way. These directions are

FIG. 179.

necessarily merely of a suggestive nature, but if the suggestions be followed up, and if the reader possess any gift for designing, he may produce many articles in the shape of mountings for various kinds of furniture, monograms for caskets, blotters, &c., in this way.

The hinges, when finished, should be washed thoroughly in hot water with washing soda, rinsed well in pure water, and, when dry, dipped in a mixture of nitric acid and sulphuric acid, equal parts, with a " pinch " of common salt added. Dip for two or three seconds, and then plunge in a large vessel of cold water to remove the acid, and dry in boxwood sawdust. This will impart a beautiful gold colour to the brass, which must be preserved by lacquering, other-wise the metal would soon lose its colour from contact with the atmosphere.

Procure some gold lacquer, and, avoiding touching the metal with the hands after colouring, give an even coat of lacquer with a soft, flat camel's hair brush. Do not mind if this coat becomes cloudy, but when dry give another, and place the work carefully in a moder-ately hot kitchen oven, or hold before a fire, when the lacquer will set bright and clear. A good enamel which will answer all the purposes of gold lacquer is that known as "Silico." It may be obtained at most ironmongers and bicycle outfitters.

The cabinet itself may be stained with Stephens' Ebony Stain, and given a " dead " or semi-polish with beeswax and turpentine ; or it may be enamelled white, which will look well.

A WALL-BRACKET FOR BRIC-À-BRAC.

FIG. 180 will give a general idea of the design for the wall-bracket, which does equally well for dining-room, drawing-room, or bedroom, for the display of old china, &c. The design shown is an easy one to carry out, but requires careful making. As such ornamental pieces of furniture are invariably painted or enamelled, the commonest wood can be used in making it, so that the actual cost of material is purely nominal ; but the columns should be of hard wood—beech, oak, mahogany, or similar woods.

The length is 2ft., width 5in., and thickness $\frac{1}{2}$in., two boards being cut out to these measurements. Having squared the edges, run a fine bead on the front and side edges of both, as in Fig. 182, by the aid of a bead-router. The columns, five in number (not including the top one, which will be mentioned later on), are 6in. long, of $\frac{7}{8}$in.-square stuff. Fig. 183 gives a correct pattern of these. Each of the sides of the square parts of columns should have a fluted

moulding to correspond with that on the boards.

The arches between the columns are in one piece, $\frac{1}{4}$in. thick, $1\frac{1}{2}$in. deep, and let into the three inner columns by cutting a slot in them $1\frac{1}{4}$in. deep, as shown in Fig. 183, the remaining $\frac{1}{4}$in. being cut off the wood to be let in. The end columns are not cut through, but a groove $\frac{3}{16}$in. deep is cut in two sides for the front and side arches (see section, Fig. 184). To get all the arches true, cut a pattern in cardboard, and mark the wood from it. Fig. 180 will show sufficiently clear the proper curve of arches.

The back supports opposite the end columns (Fig. 181) are plain wood $\frac{3}{4}$in. wide, $\frac{1}{2}$in. thick, beaded, of course. These are the only supports at the back, the remainder being open.

Fasten the five columns to the bottom board from underneath by screws, the columns fitting flush with the front edge, and the back supports in a similar manner. Then, having fitted the piece of wood constituting the arches into position, and the side ones as well, fix on the top (but only temporarily) with screws.

two grooves in the top board, into which the sides (A and B, Fig. 180) are fitted (A and B, Fig. 185), which latter is a vertical plan of the top, the dotted lines intimat-

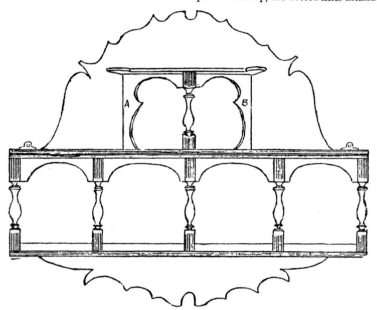

FIG. 180.—VIEW OF WALL-BRACKET.

Next proceed with the upper part, which has one column 5¼in. long, proportioned as follows: top square part, 1½in.; turned part, 2½in.; square part at bottom, 1¼in. This is fixed on the

ing the shelf. The sides A and B are of the same thickness as the arches, and the grain of the wood should be vertical;

FIG. 181.—SIDE VIEW.

FIG. 182.—SECTION OF BEAD OR MOULDING.

the side which is glued to the back must be chamfered off so as to fit close to the back, the other end being let into the column (Fig. 184). Having screwed the column to the top board, the latter can be fixed permanently, and as the screws will go in the way of the grain into the columns and uprights, see that they are sufficiently long to hold well without

centre of the top board, 1in. from the front, diagonally; but before proceeding to do this, it will be necessary to cut

splitting the wood. The top shelf is 11in. long at back, and 6in. at front, of ¼in. wood, the edges left square.

The sides A and B must be fixed into position before putting on the top shelf.

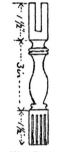

FIG. 183.—COLUMN.

The back, which should be of ⅜in. stuff, is 20in. long, 9in. high in the centre, and is fixed on the top board. The best way is to cut two pieces of moderately stout brass, 2in. by 1in., and screw them to the back of the top board, and also to the back itself; this will ensure a neat and strong job.

In cutting out the back it is better first to trace the pattern on cardboard, beginning at the top centre and working down to one end; then this reversed will ensure the other half being exactly the same. Having marked the pattern on the wood, cut out with a fret or frame-saw—the former in preference, as it leaves a cleaner edge, and the curves are more likely to be true. Fix the back to the brass plates and screw the top shelf to the back with thin screws, and also

The fancy piece at the bottom, the pattern of which is in character with the back, is 20in. long, 3in. deep, and also of ⅜in. stuff. In order to fix this firmly, screw a slip of wood, say 12in. long, ⅝in. square, to the same at back on a level with the top edge, and then screw the same under the bottom board, 1in. from the front edge of board.

On the bottom board, about 1½in. from the back edge, it is as well to fix a small square bead the whole length, so that saucers may be stood leaning against the wall, and this bead will prevent their slipping forward.

Two brass looking-glass plates, 1½in. long, fixed near each end at the back, to hang the bracket by, will complete the bracket, and the only thing to consider now is the finishing.

It should be understood that during the process of making each part should be well finished before fitting, and any

FIG. 184.—SECTION OF END COLUMN.

roughness removed by a plentiful use of glass-paper. So much depends upon small details in finishing a piece of work like this, that frequently a well-made thing is spoilt by bad finish. If, as suggested, the bracket is made of the commonest wood, all knots must be painted over by what is known as knotting to prevent their showing at all when finished, and any cracks that might happen to appear in the wood filled in with thin plaster of

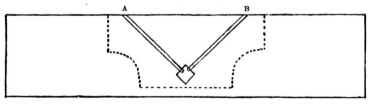

FIG. 185.—PLAN OF TOP SHELF, showing position of Column, Side Wings, and Small Shelf.

by a screw from the top into the front column. By this means the back will be kept perfectly rigid.

Paris, any of which, if left on the surface, can easily be removed when dry; then apply a coat of white lead, turps, and a little gold size to harden it. This will dry quickly, and should then be rubbed down with glass-paper. The same preparation with a little ultramarine blue mixed with it, for the second coat, will produce a fine dead white, and may be left at this stage. Some persons prefer this to enamel, and it is much less expensive; but should the amateur prefer enamel, there is no occasion for this second coat, the enamel being put on after the first coating.

A   MORESQUE   SEAT.

MANY of the quaint odds and ends of furniture which are so popular in the present day are not only useful for their primary or legitimate purpose, but also lend themselves in an accom

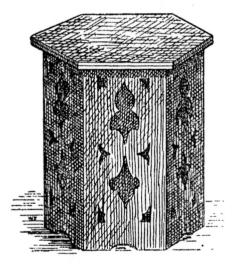

FIG. 186.—A MORESQUE SEAT.

modating way to other uses. The object shown at Fig. 186, for instance, while we call it a seat, will serve admirably to hold a small tea-tray or a flower-pot, work-basket, &c., and while performing these and other varied duties, possesses the advantage of always looking as if it were meant and made to be used for the purpose to which for the time being it is put.

But there is more here than meets the eye, for our "tea-work-table-seat" has capabilities within itself of treasuring anything that the owner of a drawing-room or boudoir may desire to stow away, e.g., a private duster or any of those hundred-and-one odds and ends (especially odds) which are always to be found somewhere in a sitting-room, but which must be kept out of sight. This is an article of furniture which is rather expensive to buy, chiefly because it has not yet come to be recognised as saleable by the manufacturers who cater for the general public. Be this as it may, it certainly will not cost any reader who is a working amateur much to make if he follows the directions given here.

Almost any clean, straight-grained wood will suit, but the choice will greatly depend upon the style of treatment with regard to finish proposed — i.e., whether the seat shall be stained, ebonised, polished, or painted, &c. Leaving the decision to the reader, we will proceed with directions for making our subject of red deal, eventually to be treated with enamel paint.

The first portion to be prepared is

represented by the horizontal shading at Fig. 187. This is a hexagon of inch stuff, and should be very carefully cut out according to the dimensions given; or, if a larger or smaller seat is to be made, the alterations must be made proportionally. A piece of inch board sufficiently large to allow of a circle 14in. in diameter being inscribed upon it is required. This piece may be formed of two widths of 9in. board carefully jointed, and joined by at least four strong dowels and glue in the usual way, then clean up true and smooth, and describe a circle of 7in. radius. Now strike out the hexagon by trisecting one semicircle and drawing lines from these points through the centre to the other side of the circumference; test these divisions with the compass, and inscribe

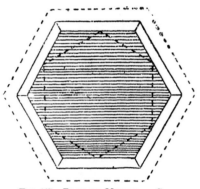

FIG. 187.—PLAN OF MORESQUE SEAT.

the hexagon, the sides of which will be as nearly as possible 7in. Cut out with a fine saw, and shoot the edges so that they shall be square to the surface and exactly corresponding with the lines as already drawn upon the wood. This hexagonal piece is to form the bottom of the seat, and is to occupy the position indicated by the dotted lines in Fig. 188; but as it regulates both the size and shape of the work in hand, we begin with it.

Fig. 188 shows the shape of the outside surface of one of the six sides; they should be of at least ¾in. stuff, 18in. long, and of section shown at Fig. 187, which will make them almost exactly 8in. wide outside, and 7in. inside

measurement. The bevelling of the edges of these sides must be very accurately done, but if the worker sets his bevel to an angle of 60deg. no difficulty will be met with. Having got

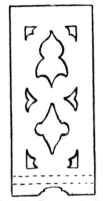

FIG. 188.—DESIGN FOR SIDE.

out the six sides, the next step will be to cut out the perforations, as indicated in Fig. 188. This may be done with a fret- or whip-saw, and, being cut, the perforations may be chamfered to about half the thickness of the board, or left plain according to fancy; the latter treatment will be more in keeping with the style of the article.

FIG. 189.—CORNER-PIECE.

Fig. 189, which is drawn upon a larger scale than the other illustrations, shows the shape and approximate dimensions of the six corner-pieces, the positions of which are indicated by dotted lines in

I

Fig. 187. These corner-pieces should next be got out; they must be of some tough wood, such as beech, and should be 1¼in. thick. The external angle must, of course, exactly correspond with the angles of the hexagon, and the external sides will be 3½in.

The six sides may now be tacked in position upon the hexagon, placing the latter about 2in. from the bottom; use thin wire nails, two to each side, but do not drive them home. When the sides are perfectly fitted around the hexagon, as shown at Fig. 187, they may be finally secured with 2½in. screws—three screws to each side—slightly counter-sinking the heads. Begin by screwing up one side; then remove the next to it, glue its bevelled edge to that of No. 1, screw on No. 2, and nail the glued corners at four equidistant points along its length, using 2in. " battens" or oval wire-anchor-heads. Next remove No. 3, proceed as before, and so on until the whole six sides are fixed in position.

The corner-pieces are now to be placed at the top of the seat, their upper surface being flush with the upper ends of the sides. These are to be fixed with glue, and two 1½in. screws on the inside, one screw being placed near the end of each corner-piece; two 2in. screws should also be inserted from the outside, one at each side of the external angle of the hexagon.

It will of course be evident that, to fulfil the requirements of such a receptacle as mentioned at the outset, the perforations in the sides of the seat must be closed in some way upon the inside. The best thing for this purpose is 4-ply Willesden Paper, but any moderately strong cardboard will suit well, or thin boards may be used; but either of the other materials will be better than wood, and if cut to reach from under the corner-pieces to the bottom of the seat will make a neat lining.

If the seat is to be ebonised or painted black, this lining should be painted any shade of red which, appearing through the perforations, will present a nice contrast, and whatever colour be selected for the seat a different but harmonising colour should be given to the lining, which should be cut out and fitted before the sides are finally fixed in position as described above, and secured with tacks when the painting, &c., has been finished.

The shape and comparative dimensions of the lid are shown by the outer dotted lines at Fig. 187. It consists of a hexagon about 2in. larger every way than that which forms the bottom of the seat; it may be of 1in. or 1¼in. stuff, and should have a simple bead running round the edge (Fig. 186). This may be made with a scratch-router, or with an American reeding tool.

The lid is to hinge upon one of the sides of the seat, and is to be provided with a small brass hasp by which it may be fastened upon the opposite side of the hexagon.

The reader can now carry out the finishing.

A SIMPLE HALL STAND.

A SIMPLE though very useful hall stand is shown in Fig. 190, as it takes up very little room, standing out but 12in., and is fitted with a good-sized drawer for clothes-brushes, a looking-glass, pegs for hats and coats, and sufficient space for a dozen umbrellas.

The uprights (A) are 6ft. 6in. long, 4in. wide, 1in. thick. The top and bottom rails (B) measure 2ft. 2in. when fitted, the same width and thickness as the uprights; wood 1in. thick is used for the whole of the back. The middle rail (C) is 6in. wide; the inside uprights are 2in. wide, mortised into the top, middle, and bottom rails, leaving a space of 5in. between them and the outer uprights, so that there will be a space 12in. wide between the middle uprights for the looking-glass. The top and bottom rails of the looking-glass frame (D) are 2in wide, and will measure 12in. long when fitted.

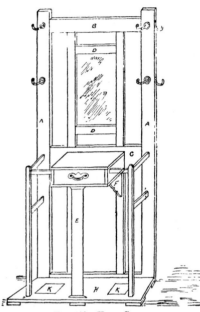

FIG. 190.—HALL STAND.

When mortising, let the length of the tenons be only half the width of the uprights and rails they are mortised into, as shown in Fig. 191; for not only will the frame be perfectly strong, but it will have a better appearance than if the tenons appear through to the outer edge.

Rebate the frame that holds the glass at the back ½in. and to a depth sufficient to let the glass in flush. This being done, the whole back of the stand can be fixed and glued together. The edges can be left square, as shown in Fig. 190, or a $\frac{5}{16}$in. bead run round gives a finish.

The box for the drawer is best made separately. When made, fix it to the middle rail by screws from behind, and support it by brackets (F) underneath. It should measure 15½in. by 5in. and 10in. deep, with ½in. wood for the bottom and sides, ¾in. for the top, and the edges slightly chamfered. Let the sides be 4in. high, and screw the top and bottom on to them. The brackets to support the box measure 7in. each way; screw these on from the back and to the bottom of box 1in. from each end. This should be done before the top of the box is fixed.

The drawer will measure 14½in. by 4in. and 10in. back to front (outside measurements), and is made in the same way as described in the article on "A Washstand," p. 71.

The bottom of the stand (H) measures 2ft. 10in. long, 11½in. wide, ¾in. thick. Cut out the parts marked K 6in. square, and 2½in. from the outside edges; and in these fit zinc trays, with flanges all round for the umbrellas to stand in. They must fit loosely, to enable them to be taken out when required.

Underneath the bottom board screw six blocks of wood, 4in. square, 1in. thick, as shown in Fig. 190. Before this is fixed, it will be better to cut the holes for the uprights of the umbrella rails. These uprights are 1in. square, the tops being

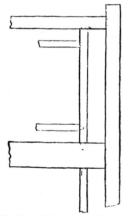

FIG. 191.—POSITION OF TENONS.

rounded off. The holes will therefore be 1in. square, and must be in a line with the middle of the uprights (A), and stand out 9in. from the back, and cut ½in. deep. The bottom can then be screwed on to the back.

The centre upright (E) is not to support the drawer-box, but to prevent any forward motion of the stand when completed; its length is about 2ft. 7in., 3in. wide, ¾in. thick. To fix it, lay the stand flat down, and screw from the bottom and through the bottom of the drawer-box (it will be necessary to leave the top of the drawer-box unfixed till this is done). Fix this upright in a line with the uprights of the umbrella rails, which stand out 9in. from the back, and fix a small moulding on at the top and bottom, as shown in Fig. 190.

The uprights for the umbrella rails are 3ft. long when fixed, the side rails (two each side) 8½in. long, the front rails 7in. Cut these out of a piece of ¾in. wood, 1in. wide, chamfer the top edges, and let these into the back and uprights and also to the sides of the drawer-box ½in., and if fitted tightly and glued will be found quite firm, so that these rails must be cut 1in. longer than the lengths given.

Let the silvered glass have a wood back to it, ¼in. or ⅜in. thick, screwed on. Put in the glass last, when the stand has been stained, for the one described can be made of ordinary deal or yellow pine.

After well sand papering the stand, stain with dark oak stain, which will almost entirely disappear in the wood, leaving it to all appearances very rough.

Sand-paper it a second time, and then apply a second coat of stain. When this is dry, rub the surface over with boiled linseed oil, and the wood will look almost equal to walnut, and far better than if it were varnished; but should it be preferred varnished, put on a coat of size after the second coat of stain, then sand-paper and varnish. So much depends upon the finishing, that it should not be hurried over, and a free use of sand paper will always repay the amateur for his trouble. First, a coarse quality to remove any little defects left by the plane, or pencil marks, then a fine quality paper to give the wood a smooth surface. Jackson's powder stain, which is prepared by adding hot water according to the instructions given with it, is a good stain for the purpose. Either a dark or light stain can be obtained, according to the quantity of water added, and the wood can be varnished or left dull with but a rubbing of boiled oil.

Fix brass hat pins, as shown in Fig. 190, and an additional one can be fixed in the centre over the looking-glass. The patterns are numerous, but the amateur can suit himself at any ironmonger's. For coats, brass wardrobe hooks should be fixed on the outside edges, two each side, and the zinc trays, made cheaply enough by any zinc-worker, should be ¾in. deep, with a ¾in. flange all round.

A SIMPLE OVERMANTEL.

TO those amateurs who have already succeeded in making an overmantel, the one shown here (Fig. 192) may seem too simple to attempt; but to those who have not, it will be found quite difficult enough for a commencement.

To make an overmantel of common deal may seem absurd to many, but the one illustrated has been made of this wood, and if carefully made and stained or ebonised, it will look exceedingly well, and the cost is very slight; at the same time any other kind of wood can be used.

The overmantel measures 5ft. long, 3ft. 6in. extreme height, and, presuming it is to be made in deal, the whole frame to be of ¾in. stuff. The bottom rail is 2¼in. wide; the centre uprights 3ft. 3in long, 3in. wide ; the other uprights 2ft. high, and the same width as the bottom rail. The lengths of the uprights are extreme measurements taken from bottom of rail. Mortise the uprights into the rail, the centre or longest ones being 1ft. 3in. from either end, with the two of the shorter uprights mortised close to them on the outer sides of the long ones. In mortising the end uprights let the tenons be 1in. wide only, and 1in. from the outer edge, and, with the remaining tenons, always cut them at least ¼in. from each edge of the uprights, so that when fitted the marks of the chisel, caused when cutting out the mortises in the rail, are out of sight.

A section of an upright showing rebate and chamfer (Fig. 193), the chamfer inclining inwards ¼in., is here given Make the columns on the centre uprights, which support the centre shelf, separately, and screw them on to the uprights from the back. The columns are 3ft. extreme length, 2¼in. wide, and 1in. thick ; make the top and bottom pieces separately and screw on to the columns.

Fig. 194 shows the section of a column. The centre pieces are of ⅜in. wood, 2in. wide, fixed on with glue and needle-points. Cut three flutes in each column in the wood by means of a " bead-router." If the amateur possesses carving tools, the flutes can be

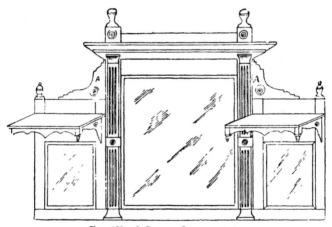

FIG 192.—A SIMPLE OVERMANTEL.

The cross-pieces between the uprights above the glasses are 9in. wide, mortised into the uprights ; the tenons are 1in. long, so as not to show on the outer edges of uprights—two tenons at each end of the cross-pieces, 1½in. wide and 1½in. from each end, are enough ; chamfer the inside edges of the frames, say ⅜in. If the amateur does not possess a proper chamfer-plane, mark out the size of chamfer on the front and edge in pencil first, then plane to the marks, so that when the whole frame is fitted together the chamfered edges may meet evenly. Rebate the back of the frames for the silvered glass ¼in. deep, ⅜in. wide.

made with a small gouge, first marking out the width of the flutes on the surface, and then cutting them out the way of the grain.

The tops and bottoms of the columns are 2in. deep, 2¼in. wide, where fixed on to the columns, and 3¼in. wide at the ends. Plane up a piece of wood, say 15in. long, 2in. wide, 1½in. thick ; hollow out the front surface so as to leave the wood 1in. thick at the top, to fit flush on to the columns, and leave a square edge at the bottom ½in. deep ; then cut off the four pieces 3¼in. long, and hollow out the ends in like manner, though this will more easily be done with a chisel and half-round rasp. The shelf over the centre glass is made of two boards : the bottom one 4in. wide, the top one 6in., and ½in. thick ; the ends of the bottom board are to come flush with the outside edge of the centre uprights, the top board to lap over 2in. at the front and ends, both being screwed together from the top, and the whole shelf, in addition to resting on the top of the columns, screwed from the back.

FIG. 193.—SECTION OF FRAME, SHOWING CHAMFER AND REBATE.

Next, attention should be paid to the side shelves. These measure 1ft. 3in. long, 6in. wide, and are supported by brackets to correspond in pattern to the pieces marked A in Fig. 192, and measure 5in. by 6in. On the front corners of the shelves fix small pieces of wood 1in. square, 1½in. long, and on the ends of these some small knobs, in character with the larger ones

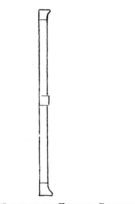

FIG 194.—SECTION OF FLUTED COLUMNS.

on the uprights, while between these small pieces on the front fix thin slips of wood cut out as shown in Fig. 192. These need only be ¼in. thick, and cut with either frame or fretsaw. To fix the corner-pieces, file off the head

of a 1in. screw, and drive the plain end into the small block, leaving about ⅜in. of the thread showing. Then screw this into the shelves, after glueing the surface as well. To fix the fancy piece at front, cut ¼in. grooves in the blocks before fixing them, and glue the slips into these. The shelves should be screwed to the frame from the back, as well as being supported by the brackets, which are fixed in like manner.

The side wings, marked A in Fig. 192, measure 8in. by 9in., cut out of ⅜in. wood. It is advisable to cut the pattern in stiff paper first, and then mark the wood out from it, by which means both wings will be alike. To fix these, screw two small glass plates on each, and so screw the plates to the frame at the back.

The rosettes on the wings and centre uprights are 2in. in diameter ; those on the centre of the columns and brackets of the small shelves, 1in. A section of these is given in Fig. 195, though it is not necessary to follow out the pattern ; and the same applies to the knobs on the uprights and under the side shelves, but be careful that they are all in character. For the uprights, use one knob of each size cut in halves ; this will allow of the overmantel fitting close to the wall. The glasses should

FIG. 195.—SECTION OF ROSETTES.

be backed with thin wood, and if the cost is not an object, bevelled plate-glass is recommended.

Now comes the question of finish. Black and gold may suit some tastes. For this use French polish and vegetable black (mixed), applied with a brush ; two coats making an excellent dead-black. Use sand-paper after the first coat to make a smooth surface before applying the second, and then use Judson's gold paint for decorating, finally putting on a coat of thin oil or spirit varnish. Should the amateur prefer it, boiled linseed oil, rubbed in with a pad, may be used instead of varnish. This will give the appearance of ebony; but the linseed must be rubbed in before putting on the gold.* In the event of varnish being preferred, the wood must be sized after being blacked. For imitation walnut, use two coats of dark oak powder stain (a shilling tin making a pint of stain), finishing with either boiled oil or size and varnish. Stephens's or Olympia stains may also be used.

COMBINATION OVERMANTEL AND CIGAR CABINET.

THE illustration (Fig. 196) shows a more advanced design for an overmantel than the one previously given, the use of the lathe being brought more into requisition, though there is nothing difficult about the whole work; it only pre-arrangement is a common error amongst amateurs, and many a piece of work begun carefully is hurriedly finished, and the otherwise good effect spoilt.

First, with regard to the material

FIG. 196.—COMBINATION OVERMANTEL AND CIGAR CABINET.

wants to be carefully thought out before commencing, and not to be hurriedly finished. The want of this

with which to make it. Anything from ordinary deal upwards will do, only in the event of deal being used the columns must be of a harder wood, say beech. In this case the overmantel should be painted with some delicate tint of

*A good " dull finish " is to be obtained from the " Egg-Shell Gloss " polish, sold by the Torbay Paint Co.—ED.

enamel, or ebonised and relieved with gold, or treated with oak or walnut stain, as recommended with the design previously given.

The whole length of the overmantel is 5ft., and the extreme height 4ft. 3in. The length of course can be made to suit the mantel—for a small room, 4ft. 6in. would be long enough. However, in describing, 5ft. is assumed to be the length.

Fig. 196 shows the overmantel with a plain back. This has a good effect when veneer is used and the whole surface well French-polished. Let the back be as shown in Fig. 197, the inside edges of the frames, with the exception of the centre one where the glass goes,

For the purposes of the description we shall describe this overmantel as being made of yellow pine, which is cheap, easily worked, and obtainable nearly, or entirely, free from knots.

The bottom rail is 5ft. long, 3in. wide, and ⅜in. thick. The end uprights are 3ft. 3in. long when fitted (extreme length from bottom); the centre uprights are 3ft. 9in. All uprights and cross-pieces are 3in. wide and ⅜in. thick. The space between the longest uprights is about 1ft. 9in., and between them and the end ones about 13in. This, of course, will depend upon the length of the overmantel; however, these are proportions that will guide the amateur.

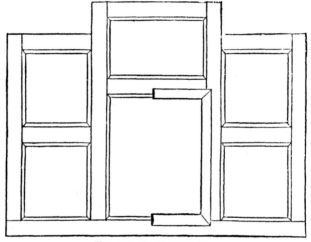

FIG. 197.—PLAN OF BACK.

being rebated from the front, into which fit wood panels flush, with a small wood moulding put round; this not only hides the joins but takes off the flatness of the back. These panels can readily be substituted by Lincrusta Walton, so much in vogue now for similar purposes. Presuming wood panels are fixed into the frame, a rich coloured plush stretched over the panel and then fixed in the frame would be very effective for a dining-room, or for a drawing-room some delicate tint would be in character, the woodwork being painted or enamelled with a colour to harmonise with the plush.

Let all frames, except where the glass goes, have a rebate on all four sides, showing from the front. There is no need to cut a rebate on the frames, which would lessen the width on the front surface: simply screw or nail small slips of wood on the inner edges of the frames, leaving a rebate ½in. deep for the panels to fit into. The looking-glass frame will also require a rebate, only not deeper than ⅜in., or at any rate according to the thickness of the silvered plate-glass.

The glass is usually fitted in from the back, but in this case fit it from the front, and to keep it in position put a thin, half-round moulding, as shown (Fig. 197), about 1¼in. wide; fix this on after the glass is in, and with glue and fine brads well punched in. Let the moulding be as thin as possible.

Presuming wood panels are to constitute the back, let them be ½in. thick, which is the depth of the fillet or rebate, and they will come flush with the front surface; then, as we have said, brad on a small moulding ½in. wide, which will not only relieve the back, but will hide any imperfections in the event of the panels not fitting well in the frames.

The front part must be added piece by piece. The bottom shelves (A in Fig. 196) are raised 3in. from the mantelshelf; they measure about 15in. long, and stand out 5in., 6in., or 7in., according to the width of the mantelshelf, the columns either being fixed on the shelf, or as drawn. Make these in two lengths, as the difficulty would be in turning in one length columns, say 3ft. long. Turn the columns out of hard wood 1in. square, and it is not necessary to follow any particular design, though, as a guide, one is given in Fig. 198.

The lower columns are about 1ft. 8in. long, the top ones on the outside being the same length. The inner top columns extend higher by 6in., and can be made in either one or two pieces. It will be noticed that the inner top columns vary in design from the others. The distance between the shelves (B) and the first moulding is 12in. In the column leave half of this square, and the front and one side fluted. These columns extend to the top centre cornice, and can be in one or two pieces, the break being made at the first moulding, as will be seen by referring to Fig. 199, showing section of overmantel with a shelf fixed, and the moulding bradded to its edges.

On the top of the outer columns, running the whole length of the overmantel, fix a board, ½in. thick: this will keep the columns firm, and form a shelf for either side, where a vase might be placed, if desired, and at the same time it forms a bottom for the cigar-cabinet over the glass. To the edge of this shelf at each side brad on the top moulding, which should be 2in. wide. The lower moulding should be 1½in., and by referring to Fig. 196 it will be seen that it extends inwards some 2in.

more than the top one. Mitre the mouldings at the corners.

The front of the cabinet can be made out of one piece if the amateur possesses a substantial fretwork machine to cut out the doors, or else it may be made up in the usual way by mortising the uprights into the top and bottom rails, the uprights being 2in. wide, and the doors, which are hung from the top, being of sufficient width to allow of a cigar-box placed inside.

The top cornice rises in the centre 7in. above the cabinet; it is of ⅝in. wood, the moulding (2in. wide) being fixed to

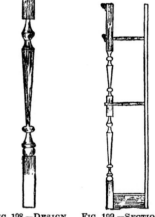

FIG. 198.—DESIGN FOR COLUMN.   FIG. 199.—SECTION OF OVERMANTEL.

the top. The wings over the sides are of ¼in. wood, fixed to wooden blocks glued or screwed on to the top board at the back. The cabinet must also have a top board to keep out dust, and the cornice-board is also fixed to this.

Between the two mouldings on the sides are fitted panels; cut these out in fretwork, as shown, or leave them plain, as preferred; this applies to the ends as well as to the front, and they should be fitted before the mouldings are fixed.

The handles for the cabinet doors are of mediæval pattern with drop rings.

The finishing and colouring must be left to the amateur, as they are a matter of taste. One thing, however, we would say, take care to finish off the overmantel well, and do not be sparing with the sand-paper.

A PORTABLE BOOKCASE.

THE drawing (Fig. 200) shows a portable bookcase shut; Fig. 201, the same open for use. In consequence of the weight of books, it is very necessary that the case should be made strong

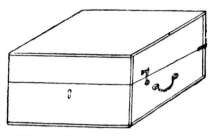

FIG. 200.—PORTABLE BOOKCASE (SHUT).

—not less than ¾in. wood, which, when planed both sides, will reduce to ⅝in. full. Taking it as shown in Fig. 200, the case measures 24in. long, 12¼in. high, 17½in. wide (inside measurements). Choice of wood is left to the amateur, though the one described is made of yellow deal, stained and varnished. The case is to be made whole first—that is to say, all four sides and ends are fastened together and sawn apart afterwards, as shown further on.

Make all sides and ends of two boards joined, and cut the boards full long, cutting to the required length when joined. In glueing the boards together, see that the glue is not too thick; and it is well to warm the edges to be glued before putting it on, otherwise before both edges are ready to be fixed together it will become partly hard, and so prevent a neat join.

In order to warm the edges, dip the brush first into the hot water which is in the outer part of the glue-pot, and brush the edges with it; then apply the glue, not thickly, but at the same time see that the whole surface of the edges is covered; then rub together the edges

glued, to exclude any air, and also to rub the glue well into the wood; then clamp the boards tightly together till the glue is quite hard. Having done one side, prepare a second whilst the first is in the clamps, and so on until all the sides are jointed.

When the boards are all jointed, cut the ends to the measurements given—viz., 17½in. long, 12½in. wide; the top and bottom, 26½in. long, 19in. wide; the front and back, 26½in. long, 12½in. wide.

First fix the top and bottom to the ends, there being left a ¾in. margin at the front and back, the ends of the top and bottom boards to be fixed flush with the ends themselves; the sides will then fit in between the top and bottom, and also flush with the ends. To fix the boards, 1½in. brads will do, put about 2in. apart, the heads punched in below the surface of the wood, and puttied over.

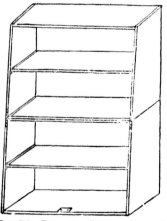

FIG. 201.—PORTABLE BOOKCASE (OPEN).

The case being fixed together, it has to be cut in halves on the slant   The depth of the box, as stated, is 12½in. ; the front must be sawn through 7½in. from the bottom, the back exactly in the middle. Mark it out carefully in pencil, then saw with the hand-saw. When cut through, plane the edges to efface the saw marks, then hinge the case at the back, where the depth of each part is 6¼in

The hinges should be of cast brass— called strap hinges or card-table hinges ; let them in flush on each half, the knuckles of the hinges protruding slightly when the case is closed. On the sides near the front edge fix brass hooks and eyes (strong), and on the bottom half a brass drop handle on either side.

The case may be fitted with a brass box-lock, say 3in. long.

Inside each half, midway between the sides, fix a shelf, or rather make it to slide in between two slips of wood, so that at any time they could be removed for travelling. The front edges of the shelves should be slightly bevelled in proportion to the sides of the case.

If made of yellow deal, the case should be stained, sized, and varnished (two coats), or painted, as preferred. Be careful that the hinges are strong; the case being a heavy one, and the hinges being fixed only on the outside edges, they are subject to a good deal of strain.

A GARDEN SWING.

THE accompanying illustration (Fig. 202) will give a general idea of a good simple swing, and also show the construction below ground which is necessary to keep the uprights from shaking loose by the action of swinging.

Deal quartering should be used throughout, the uprights and top bar being 5in. by 3in. A fair average size for a swing is 8ft. high from ground and 5ft. wide between the posts ; the top bar should overlap the uprights 3in. each side and be mortised on to them. The size of the wood will allow of a tenon 3in. square. Wedge the tenons tightly into the mortices, and then fix them still more by a stout screw or nail through the top bar and tenon from the front side.

At the bottom of the uprights fix a piece of wood, 4ft. long, 3in. square, and then fix the stays (as shown) back and front. These are of 3in. quartering, 2in. thick. Cut a notch into the uprights and bottom pieces for the stays to fit into, and further fix them with iron bolts and nuts.

The uprights should be let into the ground 2ft. deep, which must be allowed for. The stays are about

3ft. 6in. long, the tops appearing above ground. The parts underground should

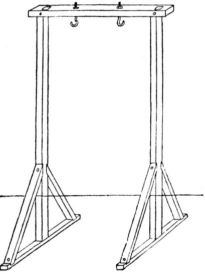

FIG. 202.—GARDEN SWING.

be well tarred, to prevent decay, and the wood should be planed over with a jack-plane before fitting, and the edges taken off, and after the swing is fixed it should be painted with at least two coats.

Proper swing-hooks are made and sold at the ironmonger's; these should go through the top bar, and be fixed at the top with a nut; but care must be taken to have a good-sized iron washer between the nut and the wood.

We must also point out the necessity of having galvanised iron eyelets to work in the hooks; over these eyelets the rope is placed, and thus accidents arising from the rope being worn through by friction are entirely avoided. Prob able cost of swing as shown, with hooks, but exclusive of painting and tarring, or concrete, say 12s. The line drawn through is the ground line. After dig ging the holes for the uprights, the bottom should be filled with stones or broken bricks, to make a firm founda tion. Or, better still, if after placing the feet of the swing in the holes, the holes are filled up with solid concrete, made of portland cement and clean coarse sand in equal parts, it will keep the swing perfectly firm and will preserve the wood.

COMBINED WALL-DESK AND PIGEON-HOLES.

THOSE readers who are rather cramped for room in their houses, and who are desirous of being methodical in the storage of their various papers and cor-

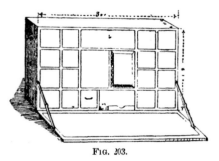

FIG. 203.

respondence, will find the article about to be described very useful and by no means difficult to make. The dimen sions and fittings can of course be varied to suit the wants and tastes of the maker, and the article may be nicely made of oak or pine, dovetailed together, and polished or varnished ; or it may be made of common deal boards, nailed together at the angles, and covered with bookbinder's cloth to hide the amateur workmanship.

The first requirement is a box 3ft. long, 21in. wide, and 11in. deep, outside measure; and of whatever material it is made, it must be squared very accurately. For the box here described use $\frac{3}{4}$in. match- boarding, and for the lid $\frac{7}{8}$in. pine. At the back three 2in. ledgers are screwed, which, besides helping to strengthen the frame, afford addi tional security for fixing to the wall.

The internal fittings and the thickness of the sides take away $3\frac{1}{4}$in. from the total length of 36in. ; thus, the sides and the two divisions on either side of the large space (Fig. 203) $\frac{5}{8}$in. each = $2\frac{1}{2}$in., and the two minor divisions of $\frac{3}{8}$in. each = $\frac{3}{4}$in. more. We thus have a clear space of $32\frac{3}{4}$in., say $32\frac{1}{2}$in., to deal with, and this may be divided thus: Top centre space, 14in.

or 14½in, or sufficient to more than take a full sheet of foolscap laid flat; the side holes in equal portions of the remainder.

To build the pigeon-holes, first cut the uprights to fit, and then fit the six cross-pieces in like manner, each piece being numbered, or marked, to ensure its being placed again in its original position when all is ready for the final fixing. Then cut the uprights half-way through with a slit the width of the thickness of the cross-pieces, as shown in Fig. 204, and the cross-pieces are to be provided with similar slits as Fig. 205. If the slot of Fig. 205 is slipped into A, Fig. 204, the pigeon-hole will be formed. The ends of the cross-pieces (Fig. 205) may be slightly let into the sides and centre uprights of the case, or they may be supported by tiny strips of wood, either glued or bradded on to the uprights. The former is the more elegant way of finishing, but the latter is much the easier method. The large central space is left to hold small account or other books. To its right is a small cupboard, and beneath these is a drawer and a vacant space for ink, pen-tray, and sundries.

Cover the writing-flap, to within a couple of inches of the edge, all round with leather or leatherette-paper, either of which can be procured at a book-binder's, and let the supports for this flap be either the usual rule-jointed, brass desk-flap holders or small chains. The advantage of the latter is that with

FIG. 204 FIG. 205.

a few spare links the flap can be fixed at any angle, which is very convenient. If the case is to be covered with book-binder's cloth or any other material, it will be best to line the divisions and internal fittings first, so that the outside covering may effectually cover up and finish off the overlappings of the edges.

A MORESQUE LAMP STAND.

MOORISH or Arabian lattice work —which in many instances is really fretwork, the Oriental title applying only to the design—is most effective for many articles of furniture, such as screens, light tea-tables, cabinets, and many other things too numerous to mention.

The suggestive sketch (Fig. 206) represents a very useful stand, especially for a small room, as it occupies little space, and will serve as a lamp-stand and receptacle for magazines, &c.

The stand may be made of almost any wood; but beech for the legs, birch for the top and shelves, and any of the harder woods usually employed in fretwork for the lattice and other pierced portions, will do well; or the lattice may be of sheet brass, which in combination with black or two or three shades of Indian red for the woodwork, will be most effective, and to anyone possessing a good fret machine it is almost as easy to cut sheet brass as wood. Let nothing induce the reader to apply ordinary fret designs to the construction of such articles of furniture as our present subject ; if this be done the character of the whole will be destroyed, and that which otherwise would be a charming addition to any room will have "common fretwork" branded upon it, whereas if a little ingenuity and pains be expended upon the cutting of designs, such as those suggested in our illustrations, the effect will be artistic and uncommon. Though not actually Moresque, they are Oriental in character.

The piece of furniture under consideration is not constructed in accordance with the ordinary rules of cabinet-making: there are no difficult joints, &c., and no turning. The cutting of the lattice is the only operation which is at all difficult, and this only to those to whom fretwork is new. To the innumerable fretworkers who possess machines it will be plain sailing, while it will have the charm of novelty for most.

The stand is 45in high, the shelves 15in. square, and the top 18in. square. The legs are 1¼in. square, and are bent or curved outwards for about one-third of their length from the ground.

The wood required is as follows. Three pieces of birch 15in. square by ¾in. thick, one piece 18in. square and ¾in. thick. All these must be perfectly square and furnished with a simple bead round the edge They must also be finished with sand-paper and well

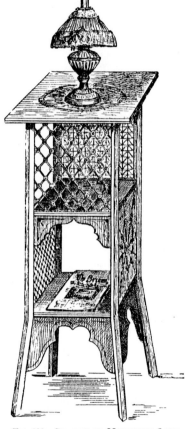

FIG. 206 —SKETCH OF MORESQUE LAMP STAND.

rubbed with a wisp of clean shavings, which will close the grain and impart a good surface. When speaking of these measurements, we refer to finished work, assuming that it is unnecessary to give particular directions as to cutting-out and cleaning up these simpler portions.

For the legs a piece of well-seasoned, clean-grained beech 4ft. long by 9in. by 1¼in. is required. Fig. 207 indicates the method of cutting them out. They must be carefully and accurately laid out with a straightedge and pencil, regulating the angles with a bevel, and using a very fine sharp saw when cutting out. First clean up the beech board on both sides, which should reduce it to the required thickness —viz., 1⅛in. True the edges so that they shall be perfectly parallel, and with the straightedge and pencil mark out the four lines longitudinal as indicated 1¼in. apart. Now take the point A 15in. from one end, and draw the line AB at an angle of 80deg. with the edge of the board. Next, from the points of intersection of the four lines with the line AB erect as many perpendiculars, draw the line CD parallel to AB, and cut out carefully. The raw sawn sides of the legs must now be cleaned with a very sharp, close-set plane, and reduced until they are 1⅛in. square, or until they are exactly the same size. The legs are now set standing together side by side with the angles corresponding, when the upper ends should be marked square and cut 30in. from the outer angle of each. There

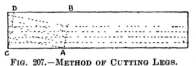

FIG. 207.—METHOD OF CUTTING LEGS.

should now be four legs precisely similar in shape, length, and substance, and the next step is to prepare the three shelves so that the legs may be attached.

Fig. 208 shows the method of fixing the legs, which consists in cutting out a piece from each corner of the shelves so as to form a rebate or bed about ⅜in. deep for the legs; these are secured by screws passing through the small corner slips (A), attached to the undersides of the shelves. They are of 1in. stuff, of birch or some other moderately tough wood, and are secured to the shelves with glue, and also two 1½in. screws. The hole in each slip, which is to take the stout 1¾in. screw to secure the legs, must be bored and countersunk before the slips are fastened to the shelves, and a long spindle screwdriver will be best for sending these home.

The margin (BB), ½in. wide, is partially filled by the lattice and small arches in the closed and open sides of the stand respectively, and the ends of the corner

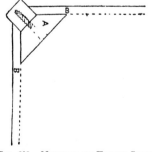

FIG. 208.—METHOD OF FIXING LEGS.

slips serve for attaching the upper sides of these. The remaining portion of the

margin—about ¼in.—shows as a bead projecting beyond or outside the lattice, &c.

Having prepared the three shelves, cut out the corner rebates and attach the slips. The legs may now be fixed, using a little glue, and screwing up tight. Great care will be required in order to accurately ascertain the position of the shelves. In the case of the upper shelf, the ends of the legs will be flush with the upper surface, in which position they should be secured; then mark off the position of the second or middle shelf, which will be about 15in. from the top, and the third shelf will be the same distance from the second.

FIG. 209.—LATTICE PANEL.

The depth of each division, however, will depend upon the height of the whole, a good proportion being to divide the stand in three equal parts, the first and second forming the spaces between the shelves, and the third being the space between the third shelf and the ground. The lower edge of the third shelf should come exactly to the angle of the legs (Fig. 206).

Next, the top leaf is fixed in position, resting upon the upper shelf, round the edge of which a little glue must be placed, and the leaf firmly secured by four inch screws, one near each corner.

It will be observed that each compartment of the stand is open upon two sides, the sides which are open in one compartment being closed in the other, therefore four square panels of lattice-work are required. The openings are also ornamented with small arches,

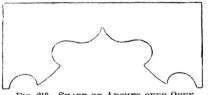

FIG. 210.—SHAPE OF ARCHES OVER OPEN SIDE SPACES.

which should be Moresque in character, and the upper portions of the spaces between the legs are also similarly treated on all four sides.

The panel of lattice represented at Fig. 209 will give an idea of what is required ; each of the four panels should be of a different design. The exigencies of space, however, prevent our giving more than one suggestive panel. We have only to repeat the warning not to employ the conventional style in fret-work designs. The same remarks apply to the arches. Figs. 210 and 211, again, are merely suggestive.

With regard to securing the lattice-work, as before pointed out, the ends of the corner slips supply two points for secure attachment upon the upper side of each ; a fine brad at each of these points will go far to give " fixity of tenure " to these light portions of the stand. The sides of the lattice should be bevelled to suit the angles of the legs, against which they rest, and to which they may be fastened with glue and a few invisible pins and very fine brads. The lower edges are secured by tacking a very light triangular slip about ½in. wide to the shelf, and glucing the lattice thereto.

As to the treatment of the stand,

whether it shall be painted or polished, &c., this must be left almost entirely to the reader's own taste, but due regard must be had to the character of the article and to the material of which it is made. We suggest that Indian red in two or three shades would suit admirably for the top, shelves, and legs, while the lattice, &c., might be bronzed, or, as before mentioned, the lattice might be cut in sheet brass and the wood-work ebonised.

To cut brass lattice, procure a sheet of brass of the required size and shape (No. 23 B.W. gauge will suit best). Place the metal between two thin boards of hard wood, as used for fretwork, and place or draw the design upon one of these. Provide against the possibility of the brass shifting its position by securing all three together firmly, and then proceed in all respects exactly as in ordinary fret-sawing. The very best saw-blades should be used—Nos. 1, 0, or 00, Griffin's patent, will be necessary.

The metal should be well polished with emery-flour and rotten-stone and oil, and washed in hot water with soap

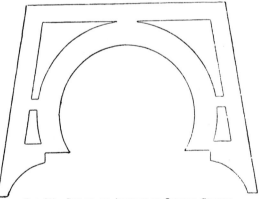

FIG. 211.—SHAPE OF ARCHES IN LOWER SPACES.

and soda *before* placing between the boards or the brass should be procured ready "buffed." When cut give a coat or two of gold lacquer, and it will be ready for fixing in position.

A REVOLVING BOOKSTAND.

THE total height is 4ft. The top is square, measuring 2ft. on each side, and, as will be seen, is tolerably massive looking. The shelves are a trifle less than the top, and may at first sight appear flimsy-looking, being only some ½in. thick; but from the mode of fitting together, they are strong enough for any weight of books that can be loaded on them. The thin rails on the shelves, as well as the bars extending from top to bottom, not only keep the volumes in their places but are important factors in the rigidity of the stand. The centre-block is hollow, and contains the support on which the whole revolves.

With one unimportant exception, and of course nails, no metal is required in the stand described; it is essentially a wood-worker's job. Let the wood be thoroughly dry — not merely thoroughly seasoned, but dry. If the wood be stood in a warm, dry place for a few days before working it, it does not shrink after being made up.

No one will begin to work without having first prepared a working drawing, or, at any rate, a full-

FIG. 212 —REVOLVING BOOKSTAND.

sized "setting out," from which all the parts can be accurately measured. We shall not give all dimensions here, but shall regard our model as serving to illustrate principles rather than details, its size and arrangement being capable of considerable variety as occasion may require. For example, there may be only one shelf instead of two, or there may be three or more between top and bottom, with only a few inches between each.

Begin with the central pillar, or, rather, with the outer casing which contains it. The pillar itself can be left for after consideration. The casing, which measures 5in. on each face, is merely a square tube composed of four pieces of inch board strongly fastened together. Make this casing of pine, and afterwards veneer it to match the remainder of the article, which will be of mahogany, walnut, or ash. The veneer may easily be laid with a veneering-hammer or a caul. Fill in the top end of the casing to a depth of, say, 1ft., with a solid block of wood, or, what will answer the same purpose, close it with a stout piece of board.

K

The former is the method generally adopted, but provided there is at the end a block or flat surface on which a metal plate can be fixed for a pivot on

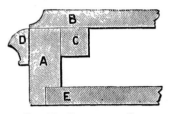

FIG. 213.—SECTION OF TOP.

the central solid pillar or axis to work against, nothing more is necessary. This block, whatever its length, must exactly fit into the tube. The pivot is of the simplest character, being only a plate of iron—say, about ¼in. thick—and a piece of ½in. iron rod, 3in. or 4in. long, with a thread turned on it at one end and either pointed or rounded off smoothly at the other, which works in a hole sunk in the plate to fit it. Fix the plate to the lower end of the block, taking care that the hole or socket within which the iron pin will fit is exactly in the centre, or the stand will not revolve properly. The block can easily be fixed in, either while the case is being made, with glue and a brad or two, or afterwards; any nails, of course, being driven in before veneering.

We may next consider the top, the arrangement of which will be understood from Fig. 212. If preferred, a plain solid top, not over 1in. in thickness, will be equally useful, but not so effective-looking. As shown in Fig. 213, in section, it consists of the square framing A, on top of which, and secured to it by the block C, rests the top B. The moulding D is planted on the face of the framing, and the bottom piece E is simply for the purpose of enclosing the whole of this upper part, and at the same time holding the axis casing in position, as will be seen by reference to Fig. 214, where the different parts just mentioned will easily be recognised.

The stand being square, it follows that each side of the framing (A) must be of the same length, viz., 2ft. The size of the shelves will be regulated by this

framing, for the upright laths are fastened both to it and to them, as well as to the bottom board, which is shown by F, on Fig. 216. If preferred, the size of the shelves may be set out first. Form the framing F of 1in. stuff, 2½in. wide. If made of solid wood, the strongest and neatest way will be to fasten the corners with the mitre dovetail, which will prevent any end grain being seen; but plain mitres, keyed together by slips of veneer in the ordinary manner, and strengthened inside by short perpendicular blocks, glued in, will answer just as well. Well fasten on the moulding D with good glue, so that there is no risk of it falling away. If there should be any fear of this, a brad or two may be driven through under the hollow, where the punched-in nails may be filled up with stopping.

The piece B is ½in. stuff. Cut this accurately to the size of A, work a hollow round the upper edge, and glue it down to A, strengthening the joint by blocks (C) glued in at intervals. E may

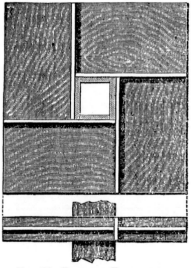

FIG. 214.—PLAN AND ELEVATION OF SHELVES

be of the same thickness as B, and as it will hardly be seen when the stand is made, a piece of pine will do very well for it. As shown in Fig. 213, let it be sunk in a rabbet in the framing, which, of course, is prepared before this is put together. The top of the pillar casing may rest against this piece (E); but it will be much better to let it run through to B, as shown in Fig. 213, as the whole of the top, which would otherwise depend for rigidity on the vertical side-laths only, is thereby held much more firmly. As the piece E will have to be jointed up to get the requisite width, this will entail very little extra labour. Fig. 214 represents the formation of the shelves in plan and in elevation, the central casing being shown in position, though it must be distinctly understood that it is not a structural part of the shelves, which it merely supports.

Each shelf or tier is composed of eight pieces of wood, four of which are the shelves proper, the others serving to connect them with each other and the central casing, as well as increasing their rigidity, and forming a stop for the books. Owing to the peculiar construction of the tiers, thick wood is quite unnecessary, and they will be quite stout enough if finished ½in., which in other words means ¾in. stuff. The division pieces may be of the same substance, and about 2½in. wide. The mode of connecting these parts is shown on an enlarged scale in Fig. 215, where it will be observed that the shelves are slightly grooved into the other pieces. A deep groove is not necessary, ₁₆in. being quite sufficient if the wood is dry.

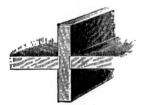

FIG. 215.—CONNECTION OF SHELVES AND DIVISIONS.

If there is any doubt about this, the rails may be of 1in. stuff, in order that the grooves may be more deeply cut, so that there may be more allowance for shrinkage. The upright bars, however, will prevent the wood from having much play. It may be as well at this point to explain how they are attached to the

central casing, though this part of the work will be best left over till the plinth or bottom is made up.

Accurately mark the casing by means of a square, and fix the rails according to these guide lines, in order that the surfaces of the four main parts of the tiers may be perfectly horizontal and on

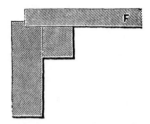

FIG. 216.—SECTION OF BOTTOM SHELF AND PLINTH.

the same level. Fix the rails by a few screws, those above the groove where they may be visible being of brass for the sake of appearance; or by brads, the heads of which are well sunk and the holes filled by stopping, or the screws may be entirely kept out of sight by running them through the grooves, as the shelves will then hide them completely. If they are inserted through the grooves it goes without saying that the heads must be well sunk in order that they may not interfere with the shelves. To fix these in it is only necessary to push one end and one side of each into the grooves corresponding, and when the upright bars are placed, these will keep the shelves in position.

The plinth is formed very much like the top framing, but is somewhat deeper, being made of stuff 3in. or 3½in. wide. It must also measure ½in. more across each face, as on the top of it rest the bottoms of the upright bars, or it would be better, perhaps, to say that these shoulder against it, for, so far as actual support is concerned, it would not matter if they were away from it altogether. Fig. 216 shows the construction of the plinth with the bottom board (or tier) resting on it, and as the plinth is ½in. more than the top framing, and the bottom shelf is the same size as the others, it follows that there is a

K 2

space of ⅛in., or a kind of step from the bottom shelf (F) to the plinth for the reception of the bars. The plinth has a shallow rebate cut round it, in which the bottom board is sunk and kept secure by as many blocks as may be considered necessary.

Attention may now be directed to what may be almost regarded as the principal feature of the stand, viz., the central pillar on which the whole revolves (Fig. 217). F is the bottom shelf, similarly lettered in Fig. 216. G is the central casing, projecting, say, about 1½in. below F, to which it is secured by blocks I; or, rather, these form almost a second casing outside the other, so that the piece H, which has not been previously referred to, rests against them as well as against the ends

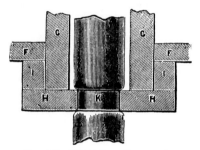

FIG. 217.—FITTING AT BOTTOM OF COLUMN.

of G. The central block, on which the whole revolves, will have been fitted with the iron pin, which revolves in the socket-plate already mentioned.

Insert the central column and mark on it just level with the bottom of the casing. Now take the thickness of H— which we may assume is ¾in., that being a suitable substance—and mark that off from the other line in the column. Between these two lines the column must be turned away (K) to the depth of, say, ¼in., the great thing to observe being that the part is truly round and smooth, for, as will be seen, the accurate turning of the stand depends as much on this part as on the iron pivot above. By slightly rounding off the shoulders left, by reducing the thickness of the block at this part, there will be less tendency to bind, for they

bear on the piece, or rather pieces, H, which may next have attention.

Fig. 218 shows this bottom piece or capping, which not only binds the

FIG. 218.—BOTTOM PIECE OR CAPPING.

central column in position, but acts as a bearing, and shares the weight of the whole of the revolving part of the stand with the iron pivot above. It should be of some hard, strong wood, such as oak or ash, and, as has already been stated, about ¾in. thick. The size of the piece should be at least sufficient to cover the casing (G, Fig. 217), and if it overlaps I so much the better, as additional strength will be gained. Cut a circular piece just the diameter of the thickness of the turned part (K) from the centre of it, as shown, and saw the board across. It is then only necessary to place the column in position in the casing, and screw these pieces on to the end of G, when the central pillar

FIG. 219.—BASE.

should revolve freely without binding. The column may be turned throughout its entire length; but as it need not be finely finished, it will do just as well if it has the corners taken off, making it in fact octagonal in section, and a good deal of unnecessary labour will be thereby saved. The dotted circle in Fig. 218 surrounding the hole represents the relative position of the shoulders of the column.

The feet or support of the stand may next be formed and fitted. The central column projects some five or six inches below the bottom of the plinth, as it is to be fitted into the base. This consists merely of two pieces of 3in. by 3in. wood, long enough to just extend beyond the plinth, as represented in Fig. 219.

Join these pieces together in the centre, where they may be secured by a nail or two. It is then only necessary

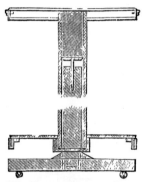

FIG. 220.—CONNECTIONS OF TOP AND BOTTOM WITH CENTRE COLUMN.

to cut a tenon on the bottom of the column to fit through a corresponding mortise in the base, and fit them together, securing the joint firmly by means of a wedge driven in from below. The stability may be further insured by

the application of blocks glued to the column and the + base, as indicated by the triangular pieces prominently shown on Fig. 220. Fit strong plate casters, as indicated, one near each end of the base pieces.

It now merely remains to fit the shelves on as already described, and

FIG. 221.   FIG. 222.

SECTIONS OF LATHS.

then to secure them by the upright bars seen in Fig. 212. For a stand of the size described twenty of these bars will be required, arranged as shown. Wood ¼in. thick will do very well for them, and they may be about 1¼in. in width. The length should be such that they will extend from the under edge of the moulding of the top to the upper edge of the plinth, so that the ends may be respectively fastened to A (Fig. 213) and F (Fig. 216). Use screws to fasten them both to these and the shelves, and for the sake of appearance the round-headed brass variety will be best. Very often however, brads are used instead, and concealed by upholsterers' plain brass studs driven in close over them. The shanks of these, however, are so weak that it is necessary to bore holes for them with a fine bradawl. Where the upright laths cover the partitions in the shelves use three screws, one in the shelf and two in the partitions. The edges of the laths may be left square, but it is usual either to run a bead along each or to bevel them off, the former method being the better-looking of the two. Both forms are shown in section in Figs. 221 and 222.

AN OVER-DOOR.

THIS item of furniture is not so popular as it should be, especially in rooms where economy of space is a *desideratum*. We have over-mantels, brackets, and cabinets, to any extent,

but comparatively few "over-doors." The present example is of the very simplest kind, but if made on the lines laid down, and enamelled in white or some pretty shade of colour to match or contrast with its surroundings, it will well repay the maker, and will serve as a useful and effective addition to a small drawing-room.

Almost any sound wood will suit, and the quantity will, of course, depend upon the size of the door, room, &c. For a door the outside casing of which is 4ft. wide, the extreme width of the over-door will be about 4ft. 9in., and its height 2ft.

The back, being the most important

workman possess neither of these appliances, the edges may be left plain and square. The wood employed, say red deal, should be 1in. thick, clean, and straight-grained.

The best method of making the back is that indicated at Fig. 224, which consists in joining two triangles with the grain running as represented in the cut. The object of this is that in the portion in which the scrolls are cut the grain runs longitudinally. Otherwise

FIG. 223.—FRONT VIEW OF OVER-DOOR.

part, comes first. Fig. 223 will give the reader an idea of its shape. The outline should be a combination of free curves, and if the sketch given in Fig. 223 be enlarged in due proportion the result will be found satisfactory.

The back may be made perfectly plain, or it may be improved by having a double or treble simple bead run all round upon the face of the edges. This is done with a scratch-router, or with an American reeding tool; but should the

the edges would consist of grain on end, which is unsuitable. When made in the way directed, the curves are more easily cut, the weaker portion is less liable to split, and when the shelves and brackets are attached the whole will be perfectly rigid and free from danger of warping, &c. The joining of these triangles (A and B, Fig. 224) requires accurate work, but if the joint is carefully trued up, the edges glued and dowelled together, and a temporary stay tacked across the back, the shaping may be proceeded with without danger. Clean up the surface well, and finish with fine glass-paper and a rub of wood-shavings ; this will prepare the way for

latter), the heads in either case being slightly sunk and covered with putty after the first coat of paint has been given.

The over-door is now complete, as far as construction is concerned ; but

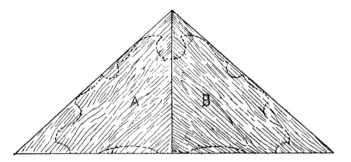

FIG. 224.—METHOD OF JOINING TRIANGLES FOR BACK.

an even coat of white enamel, which should be the next step.

Fig. 226 shows a side elevation of the over-door, the portion shaded representing the door-casing upon which it rests.

The shelves are of ¾in. stuff, and in shape should be something like that suggested at Fig. 225, but, of course, varying in depth and length—the top shelf being about 16in. by 6in., the middle shelf 36in. by 8in., and the lowest shelf 54in. by 7in. Treat the edges of the shelves in the same way as those of the back.

Fig. 226 also indicates the shape of the brackets, while Fig. 223 shows their respective positions. The lower bracket

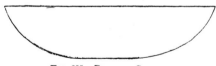

FIG. 225.—PLAN OF SHELF.

must be made to suit the projection of the door-casing upon which it rests. All five brackets are of ¾in. stuff, or stuff 1in. in the rough when cleaned up might, perhaps, be better. Furnish these also with a bead to match the shelves, &c.

Secure the shelves and brackets to each other and to the back either by battens or by screws (preferably the

although directions for putting together the various parts have been given, this should not be done in practice until the painting has been finished.

In the case of enamel colours, it is difficult to impart an even surface to subjects which involve corners and intricacies of any kind. Owing to the quick-drying properties of these paints, they do not permit of the repeated applications of the brush which are necessary, and which are easily accomplished with ordinary paint. I therefore strongly recommend that the work be painted in its various parts before finally putting all together ; by this means all surfaces will be smooth and even, and that enamelled effect which is so charming may be insured. Each coat — except, perhaps, the final one—should be rubbed down with the finest glass-paper, or, far better, with the finest ground pumice-stone and water, applied with a soft pad of rag. After rubbing

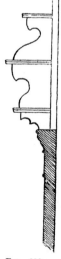

FIG. 226.—
SIDE
ELEVATION.

down until perfectly smooth, wash well in clean water to remove all trace of pumice, and rub briskly with a soft dry cloth, when, after allowing a short time for all moisture to evaporate, the next coat may be applied. Even the final coat may be rubbed down, as this takes off the glaze and imparts a beautiful eggshell-like surface ; but this is altogether a matter for taste to decide.

A WOOL-WINDER.

THERE are various forms of this useful article of a more or less elaborate kind, but usually requiring a considerable amount of humouring, and

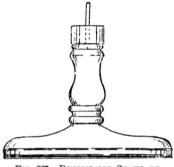

FIG. 227.—DESIGN FOR STAND OF WOOL-WINDER.

almost always possessing the attribute of ricketiness.

The winder here described is of an ancient pattern, but is a thoroughly trustworthy machine, which is not liable to get out of order, and is so substantial and durable that it may be handed down from generation to generation.

Fig. 227 is a side elevation of the stand, which may easily be made by anyone who possesses, or has access to, a simple wood-turning lathe. The height should be about 6in., the diameter of the base being 6in. also. Any hard wood will suit for this stand, and it may be made in one piece, or the pedestal and base may be distinct, joined by a screw or simple tenon. Fig. 227 also shows the position of an iron or brass

pin, which serves as a pivot upon which the arms of the winder revolve. This pin should be of $\frac{3}{16}$in. wire, about 3in. long, to allow of 1½in. being inserted in the pedestal (*see* dotted line) and 1½in. to project above. A depression circular about ½in. deep and 5in. in diameter, should be turned in the bottom of the base, into which two or three discs of ordinary sheet lead should be fitted to give weight and stability to the winder. These discs may be fastened with small wire nails or screws, and the whole bottom covered with baize. To do this, cut a piece of baize a little larger than the base, give the latter a coat of strong glue : holding it before a fire, attach the baize, and then trim round the edge with a pair of scissors. We now have a strong stand, which will be quite substantial enough to resist any force likely to be applied during the operation of winding a skein of wool or silk.

Fig. 228 shows front and Fig. 229 side elevation of the arms, two in number, of the winder. These should be about 2½ft. long, 1½in. wide at the centre, tapered to about ⅜in. at the ends, and ¾in. thick. They must be

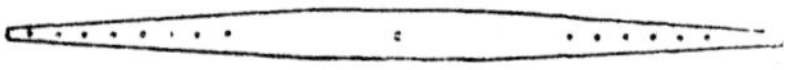

FIG. 228.—FRONT ELEVATION OF ARMS (2).

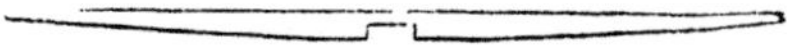

FIG. 229.—SIDE ELEVATION OF ARMS (2).

halved into each other at the centre, as shown in Fig. 229. The tenons should be carefully and cleanly cut, as the arms are thus joined together when the winder is in use, being taken asunder when not required ; therefore, whilst fitting well, the tenons must not be too tight. The arms are to be furnished with rows of holes, commencing from about 9in. from the centre, and extending, at intervals of 1in., to the points. The holes should be bored with a ¼in. bit, and are to take the upright pegs (Fig. 230), by which the wool, etc., is held in position during winding, and which may be moved to suit the sized skein in hand. Each of the four sets of holes should be numbered from the centre to the points, or *vice versâ*, and the four pegs should always be in corresponding holes in order that the wheel composed of the skein and arms may revolve smoothly.

Mahogany will suit best for the arms, but any light wood will do provided that it is straight in the grain and free from knots and flaws. If made of mahogany they should be polished, but if of pine, etc., two or three coats of Aspinall's enamel will render them quite as presentable as need be.

The pegs are to be of the shape indicated at Fig. 230, and may be of ebony or any other hard wood, or if

FIG. 230.—SHAPE OF PEGS (¼).

bone or ivory is available so much the better. They should be about 4in. long—*i.e.*, allowing 1in. for the shank to be inserted in the arm, and 3in. to carry the wool, etc.

Finally, a hole must be bored right through the centre of the arms for the pivot pin to pass through, and the winder is finished.

A LEAF-SCREEN.

THIS screen will be found useful either as a fire-shade or to fill a fireplace in summer. It is very easily made by anyone possessing a simple wood-turning lathe, and the style is novel. It is also capable of being highly embellished by hand-painting ; while if simply made of the material suggested, it will present an appearance quite respectable enough for a back parlour or spare bedroom. The pedestal, base, and claws may very easily be so modified as to afford ample scope for the art-handicraft of any readers skilled in wood-carving. The object, however, of these chapters being to supply simple directions for the construction of simple articles of furniture in the simplest way, the design represented by the accompanying illustrations will suffice, and will act as a suggestion to those who would like something more elaborate.

Fig. 231 is a front elevation of the screen. The leaf, as may

FIG. 231.—LEAF-SCREEN.

be seen, is of conventional shape, but, should the worker prefer it, he may

FIG. 232.—PEDESTAL.

of course adopt a different design—a natural ivy leaf, for example ; nor need he be confined to the leaf pattern at all, as various other forms, such as shells, etc., will be found effective for this kind of screen.

First take the pedestal (Fig. 232), which may be of almost any wood that comes handy, but beech will suit best.

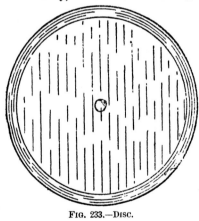

FIG. 233.—DISC.

The pedestal should be about 9in. long, and the diameter of the ball, which

forms the top, and is a socket for the insertion of the "veins," should be 3in. The thinnest part should be at least 1in. diameter, and that of the dowel ¾in. The ball is supplied with five ¾in. holes in the position indicated by dotted lines (Fig. 232), and these should

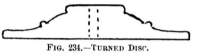

FIG. 234.—TURNED DISC.

be at least 1in. deep, being bored with a clear cutting bit, and radiating from the centre of the ball.

Fig. 233 is a disc of wood, 1¼in. thick, and ½in. in diameter, and either rounded or chamfered on the edge, according to fancy ; or this may be turned in the lathe, in which case it may be more elaborately ornamented ; in fact, anyone possessing a lathe with 4½in. or 5in. centres, or a gap bed, will greatly improve the appearance of the screen by turning this portion a little thicker than just mentioned, say, 2½in., and of some such pattern as suggested in the section (Fig. 234).

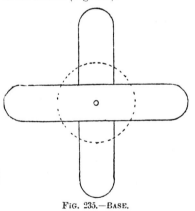

FIG. 235.—BASE.

Fig. 235 represents the base, which consists of two pieces, 15in. long, 4in. wide, and 1in. thick, rounded at the ends and halved into each other at the centre. The circular piece (Fig. 233) is placed in the position indicated by dotted lines, and is secured from the back or underneath by stout 2in. screws —say two screws through each arm of the cross which forms the base.

The pedestal is now to be fixed in position. This may be done either by simply gluing and wedging the dowel in the centre hole of the base, or by leaving the dowel long enough to protrude 1in. below the base, it being then

FIG. 236.—BALL FEET.

"tapped" with a ¾in. screw-box and fastened up with a corresponding wooden nut.

Fig. 236 shows shape of simple ball feet to be placed in each claw of the base, as shown in Fig. 231. These may also be glued in, or provided with a screw-thread in the manner just referred to.

Now comes the most important part of the screen, as it requires care and neat work—the veins of the leaf. These are five in number, and of the shape indicated at Fig. 231, the centre vein being quite straight, while the others are slightly curved. They should be of some tough, straight-grained wood, circular or slightly oval in sections, about 1½in. diameter, and tapered as represented in the illustration.

Fig. 237 shows a portion of one of these veins, from which it will be seen that the thicker end of each is provided with a round dowel or tenon, by which they are to be secured in the holes (already described) in the pedestal (Fig. 232). A perfectly straight and clean saw-cut extends from the point to within about ½in. of the shoulder of the tenon in each vein. This is to receive the material of which the leaf, etc., is composed. This latter operation calls for great care and accuracy to ensure a good job; but if the wood employed be good and tough—say good straight ash, not too dry—no great difficulty will be encountered.

The leaf itself should be of four-ply Willesden paper (ask for No. 320 Excelsior). This paper is 27in. wide, of a pretty sea-green colour, and is admirably adapted for such uses as the present. Cut out the leaf according to the design, and as large as you can make it, having regard to the size of the paper, and making provision for the ball of the pedestal. Use a very sharp knife, so as to avoid unevenness or jagged edges, and it need hardly be added that a perfectly accurate drawing in soft lead-pencil will be necessary to ensure a correct outline before attempting the cutting out of the leaf, etc.

Now fit the five veins in their proper positions without securing them, slip the leaf within the slits, and get the leaf into its place. See that the veins are at the proper angles, and that they radiate from the centre to the various points of the leaf. Having made all this sure, remove the leaf and secure the veins in their sockets with a little good glue.

When the glue is set (next day) the leaf may be again, and finally, placed in its position and secured by small pieces of copper wire passed through the veins at intervals of about 6in. along their entire length, and slightly riveted on both sides.

With regard to the treatment of the woodwork of the screen, it may be French polished or painted, according to taste; but whatever treatment be

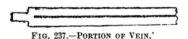

FIG. 237.—PORTION OF VEIN.

adopted for the pedestal, base, etc., the veins of the leaf should be ebonised (use Stephen's ebony stain) and French-polished; or, in case the worker possesses no skill in this direction, stain as above, and apply two coats of best clear varnish.

A CHEVAL FIRE-SCREEN.

THE frame of the fire-screen here described is of bamboo, which requires no finishing except a coat of varnish.

Fig. 238 is a front view, and Fig. 239 an end view, of the screen; the width is 22in., and the height is 1in. less have a plug of dry wood glued or pinned inside as high as the top of the sloping pieces; and the top end of each sloping piece must be hollowed to fit the upright. A hole must be bored in each foot and in the bottom of the upright for the cross-stay (as shown

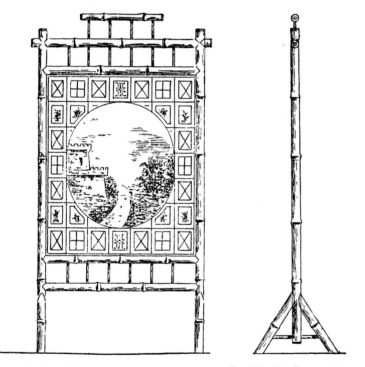

FIG. 238.—FRONT VIEW OF SCREEN.   FIG. 239.—END VIEW OF SCREEN.

than the under side of the shelf of the chimney-piece.

Procure two pieces of bamboo about 1⅛in. diameter for the uprights; fix to the lower end of each two sloping pieces for feet, and a cross-stay, ½in. diameter. The lower end of each upright must in Fig. 239), and the whole glued together, the joints being fastened with a wire nail (as in Fig. 240, which shows the manner in which all the joints are made). A screw may be used instead of the wire nail, or a hole may be bored in the upright, and the plug glued into it. The bottom of the feet must be sawn off, so that the uprights stand perfectly square from the floor.

Next get four pieces of 1½in. bamboo for the crosspieces, and plug the ends; hollow them to fit the uprights, and fix between each two of them, six pieces of ⅞in. bamboo, 3½in. long. The crosspieces must be bored, and the rails glued in. Be careful to bore the holes in line, so as not to twist them. A handle, ⅞in. diameter, and 11in. long, must be fixed on the top, with four rails 2½in. long under it, then the four crosspieces (as shown in Fig. 238). The plugs in the top piece may be allowed to project at each end, and be put through holes bored in the uprights, to receive the two projecting ends, which are 1in. long. The two uprights and these projecting ends must have a turned cap, screwed into a plug in the end of the bamboo, to cover the hole. Keep the crosspieces parallel with the floor, and square with the sides of the screen.

Eight narrow strips of bamboo must be prepared to fix the glass in the frame. The glass must be ⅛in. less than the inside of the frame. The strips of bamboo are secured with small screws, or fastened with wire nails to the frame at both sides, forming a beading round it.

If plain plate-glass or a mirror be put in the frame, they generally have a

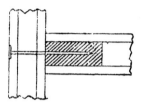

FIG. 240.—METHOD OF MAKING JOINTS.

painting done on them in oils; but the most general method is to use leaded glass. This can be had either plain or painted, and costs from 1s. 6d. per super. foot, and upwards. Sometimes needlework is inserted, in which case it will require to be stretched on a light wooden frame.

A HAND-CAMERA.

THE dimensions of the camera are 4¼in. by 3¼in. The tools required are a small tenon-saw, an iron plane, three dozen ¾in. fine brass screws, and a small screwdriver, shell-bit gimlet, and bradawl to suit them, a fine 5in. handsaw file, a sheet each of No. 0, 1, and 1½ glass-paper, and a small tack-hammer. The materials required are several lengths of well-seasoned soft wood, such as deal —(these lengths must not be less than 12in. long by 5in. in width, and ¼in. thick, or a trifle more if rough and requiring much planing); some sheet zinc (not thicker than ordinary tin-plate) sufficient to allow one dozen pieces, each 5in. by 3½in., to be cut out of it, and a few simple fittings, described farther on.

Make a long box of the wood above mentioned, to form the camera, without ends to it; the external measurement of the box must be exactly 4⅞in. by 3½in. Cut two pieces of stout wood of these dimensions, perfectly squared, and insert them loosely in the box at each end, while it is being put together.

Before screwing the box sides together they must be painted a dead black, the simplest compound being lamp-black and only just sufficient gold size to render it thin enough to apply to the surface of the wood. The sides must in the first instance only have enough screws to keep them temporarily together, as they will have to be separated again to fasten the internal woodwork.

Inside one end of the box fix a frame, against which the sensitive plates will rest, as that will be the focal plane. For this purpose strips of wood, $\frac{3}{8}$in. wide and $\frac{1}{4}$in. thick, must be fitted inside against the two sides and the bottom, great care being taken to fix them at right angles to the sides, sloping them at A as shown (Fig. 241).

The camera box must be sawn in two, and inside the front half a box of similar construction must be exactly fitted, the box being about 3in. long and projecting 2in. (*see* B on plan, Fig. 241). The hinder part will slip over this projecting part,

The velvet must be put on by gluing the $\frac{3}{8}$in. edge with thin glue, letting it remain until no longer wet but sticky. After it has been pressed on and the glue dry, the edges may be trimmed, and the strips cut and jammed firmly into place, the velvet edges being outside, that is lying next the camera, but the strips must not be fastened yet in any way, either together or to the camera.

Glue the edges of the strips facing the back, and, up-ending the camera, let it lie on the lid before mentioned until the glue holds it quite firmly,

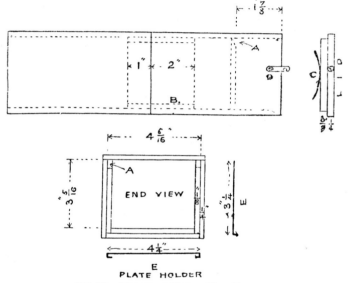

E
PLATE HOLDER
FIG. 241 —METHOD OF MAKING HAND-CAMERA.

and must be free from play, and only draw off or on smoothly, but with considerable force. The object of thus being able to lengthen the camera slightly is to provide for photographing nearer objects than usual, when, owing to their being under the control of the operator, it is safe to attempt them.

To finish the back end a lid must be cut out the exact external dimensions of the end, then a frame of strips of wood $\frac{3}{8}$in. by $\frac{1}{4}$in. must be made (*see* plan), the strips being covered on the $\frac{3}{8}$in. edge with velvet before being fitted tightly into the end of the camera.

the hand having been inserted in the back half of the camera to press the frame well against the lid. When safe to do so, the lid with the velvet-covered frame may be removed, and a pressure spring (*see* C on plan, Fig. 241) must be attached inside, as hereafter described. Two metal catches must be screwed to the end of the camera at D to hook on to screw knobs on the edge of the lid. Both catches and knobs must be countersunk.

The next step is to make the plate-holders. Twelve pieces of the zinc specified must be cut perfectly squared of such a size that when the two most remote edges are turned over, as shown at E, they will hold dry plates, 4¼in. by 3¼in. One of the long edges is likewise turned just sufficiently to prevent the plates from passing through in that direction, which will be at the bottom of the camera. To bend the zinc plates into shape at the edges, obtain a flat piece of brass or iron the exact size and thickness of the plates, lay this in the centre of the zinc, tap the edges up, and then hammer them over the metal plate, slightly turning the foot also as described. The metal plate must draw out easily, and the glass plate slide into its place, fitting without shaking.

The spring attached to the cover of the back of the camera must not be a fixture, but attached to a mill-headed screw passing through the centre of the cover or lid. A hole must be bored through the cover, slightly smaller in diameter than the screw, which must then be worked in the hole until a worm is formed in it. The end of the screw must be filed to a smaller diameter, with a shoulder, and this end must pass through a hole to be punched in the centre of the spring ; the end must then be riveted, but leaving sufficient play to admit of the spring moving freely. By turning the head of the screw outside the cover, the spring is forced forwards or backwards at pleasure, the object being to enable the operator, when he removes one of the dozen sensitive plates, to draw the spring sufficiently far back to prevent it from pressing against the remainder of the plates. The manipulation of the sensitive plates is as follows. The whole of the dozen plates having been placed in their zinc sheaths, or slides, in the dark-room, they are all slipped into the recess at the back of the camera ; the cover is then hooked on the back, and the spring forced against the hindermost sheath by turning the mill-headed screw. By raising the hindermost plate under a cover the front plate, after an exposure, may be replaced by another ; the screw is reversed, the spring drawn back, and, the camera being tipped up, the eleven other plates fall back, leaving a space in front previously occupied by the now-exposed plate, and into this space

the fresh plate is slipped with its sheath, and the spring is once more pressed against the plates by again turning the screw ; this operation being repeated until all the twelve plates are exposed.

The arrangement for thus manipulating the plates is as follows. That part

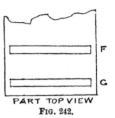

PART TOP VIEW
FIG. 242.

of the top of the camera over the recess containing the plates has two openings cut in it for the passage of each plate during changing, one opening being directly over the front plate F (Fig. 242), and the other directly over the back plate G. A lever is made and fixed in the side of the camera, to lift the back plate up through its opening ; it may be made out of a piece of brass-plate cut to a square section of ⅛in. at H (Fig. 243), hammered flat at I to admit of a nail being passed through a hole in it, to act as a fulcrum, and then hammered at J in the opposite direction to give a flat surface for the thumb when pressing it down. A light-tight hole is cut in the side of the camera (directly in a line with G, Fig. 244), the lever is inserted, and sunk in the wood-

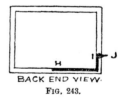

BACK END VIEW.
FIG. 243.

work of the bottom under the back plate, and a nail driven through the side at I (Fig. 243).

As this operation of plate-changing has to be carried out in the open air, it is necessary to have a loose bag of soft leather, glued, and secured with strips of wood, or zinc and small screws, over the top of the plate-recess as shown at K (Fig. 244), the plate raised by the lever being shown dotted at L, and its transference forward for the next exposure at M. The bag must be carefully examined to make sure it is light-proof before being affixed ; in order to secure this, the edges of the material must be smeared with india-rubber solution and folded over one another before sewing together. The lever being lowered by the right thumb outside the camera, it rises inside, lifting the hindermost plate sufficiently high to enable it to be grasped with the bag, which latter must be worked down, as on putting on a glove, until the plate is clear and entirely in the bag, when it is then slipped into position, as stated, at M.

For the front end of the camera, cut a piece of wood, the internal measurement of the camera, with a hole in the centre sufficiently large to allow the lens-flange to be inserted and screwed down on to it. The wood must be inserted, but not fixed, a few inches in the camera, and its permanent position at P ascertained as follows. The lens N (Fig. 244) being screwed on, the camera must be set up in view of some well-defined object 50ft. off. The plate slides being removed, a piece of ground-glass must be temporarily placed in the position previously occupied by the front slide, that is, at the focal plane, but at a distance corresponding to the thickness of the zinc from it. Covering the head with a focussing-cloth, the object 50ft. off must be observed on the ground-glass, and an assistant must push the lens (with its support) inwards or outwards until the object is in perfect focus. A large stop, say $f/8$, must be inserted in the lens during the trial. The lens must be a symmetrical doublet of 6in. to 7in. focus, covering a quarter-plate fairly well to

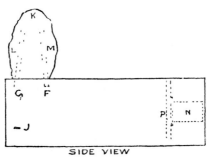

SIDE VIEW

FIG. 244.

the margin, with a stop not smaller than $f/11$. The lens support must then be fixed permanently, it having been glued just prior to the focussing test. An "Instantaneous" shutter, giving an exposure of about $\frac{1}{50}$th of a second, must be affixed behind the lens before fixing the lens support in the camera.

The cap being placed upon the lens, the camera may now be cut shorter if admissible, $\frac{3}{8}$in. being allowed beyond the extreme projection of the lens.

To complete the hand-camera put a lid or cover over the lens-end, and to attach a shutter for instantaneous work, fix a "finder," and cover the camera.

The instructions given for making the cover for the plate-end of the camera will suffice for the front. It is necessary to conceal the lens, but a hole of sufficient diameter must be cut in the lid, and a sliding cover, with a notch for the finger-nail, fixed over the hole, so that it can be temporarily withdrawn whilst an exposure is being made.

The completed camera must be covered with light-proof material, the best being the thin water-proofing used for cheap overcoats. The camera must be quickly brushed all over with hot, thin glue, and the covering drawn tightly over it.

The shutter frequently adapted to such a camera is the double-roller shutter. Two rollers stand vertically, one on each side of the lens, inside ; on one is attached a clock spring or strong elastic, and to the other a small wheel with three or four indents, and a pall falling into them, as in all clock winding arrangements. To these rollers a piece of light-proof material is fixed (such as is described for the covering is excellent if very thin), and it is first rolled round the roller with the spring. The other roller is wound up with a clock-key, and as the material winds round it, it uncoils from the spring-roller, the pall preventing it from flying back. In the centre of the material cut a hole, the full aperture of the lens, for exposures. Whilst setting this shutter by so winding it, the hole in the front lid of the camera is, of course, closed with its slide. When an exposure is made, this slide is withdrawn, and the pall released by a trigger. This can be effected with a string attached to the pall, passing through a small hole in the bottom of the camera, or by a spring trigger. A shutter that may be preferred is the revolving shutter, used in all the cameras by Lancaster and Son, of Birmingham. It has the advantage of being get-at-able, as it would be attached to the hood of the lens—the firm would supply one to any measurement—and its speed can be adjusted at will. Its trigger can be easily released by passing a wooden peg through a hole in the top of the camera, letting it simply rest on the trigger, which can then be depressed by pushing down the peg with the finger. The peg would only just protrude sufficiently.

Even the common drop-shutter could be adapted by allowing the drop to pass through slits in the top and bottom of the camera, immediately in front of the lens. These slits need not be light-tight.

The finder, which is absolutely necessary, consists of a diminutive camera-obscura attached to the hand-camera, by looking down into which the objects can be seen that are directly in front of the lens, and will appear on the negative. Briefly it is made by reflecting upon a piece of mirror, placed exactly at an angle of 45deg., the picture through any very short-focussed single lens—a common No. 10 sight spectacle lens will do—fixed vertically, and directly towards the centre of the mirror. The picture being thus reflected upwards, vertically, the finder must have a piece of finely-ground glass fixed horizontally, and by looking upon this the proper moment for making an exposure is ascertained. The finder is attached inside the front of the camera in one corner. A finder such as described can be purchased from Adams and Company, 81, Aldersgate Street, London, E.C., for two or three shillings. The first and only adjustment of the finder consists of fixing upon an object which is in focus in the camera, as already described for the adjustment of the camera-lens, care being taken that the object is exactly in the centre of the ground-glass. Then, without moving the camera, the finder must be adjusted and affixed so that the object on the ground-glass is exactly in the centre of the ground-glass of the finder.

INSTANTANEOUS SHUTTERS.

TAKE a piece of small india-rubber tubing—which can be obtained from a chemist—about ¼in. in diameter and 2ft. long, and tie one end of it firmly with a piece of fine string about ⅛in. from the end, so that no air can escape at that end. Procure from the chemist a small india-rubber ball, fitted with a nozzle, and push the other end of the tube over the nozzle to form an air-tight joint. If the joint is not air-tight it can be made so with Prout's Elastic Glue, which is melted over it by heating the glue in the flame of a candle or spirit-lamp, just as sealing-wax is melted. When cold it will be firmly glued to the nozzle. This forms the Pneumatic Release, and has to be fitted to the shutter. This is done by taking a piece of sheet zinc or thin

L

brass, about 2in. long and ¼in. wide and fastening it to the back of the shutter, by means of small screws or solder, in such a position that the end of it projects above the arm of the lever which releases the shutter. This projecting end is bent over until the bent portion is within ¼in. distance of the arm of the lever, and the end of the tube which has been tied is inserted between the bent portion and the lever, as shown in Fig. 245. It is then fastened to the bent zinc by means of elastic glue, and the zinc must be bent until it is so adjusted that the tube is pressed rather tightly between it and the lever. Now, pressing the ball forces the air into the tube, which, expanding, forces down the lever and releases the shutter.

A simple drop shutter can be made at a trifling expenditure of trouble and money. Take a piece of thin mahogany or ebonite, 4in. long by 2½in. broad by ⅛in. thick, to form the groundwork of the shutter; take two small slips of the same material, ¼in. broad, 4in. long, and ⅛in. thick, well planed, and glue them firmly down the opposite sides of the ground piece. On the top of these

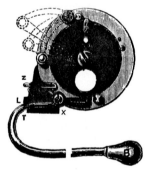

FIG. 245.—PNEUMATIC RELEASE FOR LAN-
CASTER'S SHUTTER.

T, Tube; B, Ball affixed to end of tube; L, Lever which releases the shutter; Z, Back portion of zinc or brass; X, Bent portion of ditto.

glue two more strips of the same thickness and length, but ⅜in. broad. When these pieces are glued together the effect will be to form a raised groove running down each side of the ground piece.

The next thing required is the movable piece by which the exposure is made. Take a piece of mahogany, 7in. long, 2in. broad, and ⅛in. thick, and file away the edges of the sides

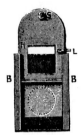

FIG. 246.—SIMPLE DROP SHUTTER BEFORE
EXPOSURE.

S, Screw at top of movable piece; H, Hole (square) in ditto, B B, Crosspiece to check descending screw; L, Lever to release shutter, the dotted line shows the positions of the lens behind the shutter.

very carefully until it is just so thin at the edges that when put into the groove and held vertically it falls through instantly of its own weight. The lens must now be taken, and a hole cut about ½in. from the bottom of the back piece, precisely the size to fit tightly round the hood, so that no cracks are left for light to enter. Mark the movable piece into three equal portions by ruling two lines across its breadth, and between the two lines cut a square hole the same size as, and opposite to, the hole cut to fit the hood of the lens. When the movable piece is placed in the grooves so that the bottom is ½in. from the bottom of the back piece, the lens will be covered. as in Fig. 246, and as it falls the holes come opposite one another, and the exposure is made, the lens being instantly covered again when the movable piece has fallen 2½in. lower. To prevent it from falling lower than this, a thin piece of wood, 1¾in. long and ¼in. broad, must be screwed by its opposite ends to the sides of the shutter (as shown at BB, Fig. 246), just above the hole made for the hood of the lens. Put a screw into the movable piece at S, about ¼in. from the top, so as to project above the grooves at the side, and when the shutter has descended sufficiently far the screw will be stopped by the cross-bar. To liberate the movable piece at a given moment, fasten a piece of thin brass 1in. long and ¼in. broad, by a screw to the top of the back piece. A very smart notch must be cut in the side of the descending piece when it is at such a height that the lens is covered completely by the portion below the hole, into this notch the thin piece of brass slips and so holds it up. On moving the long arm (L) of the brass lever the movable piece is detached, and the exposure is instantly effected. The speed of the shutter may be enormously increased by fixing two small screws or nails into the side pieces, just above B B, Fig. 246, to project above the wood. If a small elastic band is now placed over the heads of these screws, and over the screw on the descending piece, when the latter is at its full height the band will be stretched very tightly, and will pull down the movable piece with considerable force. Two or more bands may be used if the wood will stand it; but if the speed is too great, the screw at s will split the wood on the movable piece by jarring violently against the crossbar. Fig. 246 shows the position of the movable piece before exposure, and Fig 247 its position after the exposure has taken place.

With shutters of this type the best possible work can be done.

FIG. 247.—SIMPLE DROP SHUTTER AFTER EXPOSURE.

Another species of shutter which is still simpler in construction is that known as the "go-and-return" shutter. It is a very good make for small

lenses, but for large ones would be likely to cause vibration. It is constructed in a similar manner to that just described as far as the grooves are concerned; but the movable piece instead of being 7in. long is only 2in., thus saving a great deal of space. The back piece and side grooves must be about 5in. long, but the same breadth. Drive two pieces of stout wire shaped thus :

into the side grooves in the middle up to the loop. Insert a screw in the centre of the movable piece and a smaller screw beneath it about ¼in. from the bottom of the movable piece.

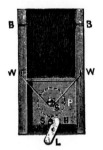

FIG. 248.—SIMPLE "GO AND RETURN" SHUTTER SET FOR WORKING.

P, Movable piece; B B, Elastic band against which the movable piece rebounds; W W, Loops for second elastic band, which is seen stretched over the screw S in the movable piece P, as set ready for use; L, Lever of brass piece H, which retains the shutter till the moment for exposure by catching the lower screw S'; S, Centre screw, over which elastic band is stretched. Position of lens shown by dotted line.

Procure two elastic bands to supply the motive power; place one round the upper part of the frame, as shown at B B, Fig. 248; insert the other in the two wire loops, which must be small enough to be stretched when in that position. Fasten a piece of mahogany 1⅞in. long and ¼in. broad at the bottom of the back piece to support the movable piece at such a height that the hole for the lens hood (which is made exactly as before) is covered by the movable piece when supported by it. On this

L 2

crosspiece screw tightly a piece of brass 1½in. long by ⅜in. broad, shaped like a hook, Fig. 249 ; this must be fastened to the centre of the crosspiece at such a height that the bottom small screw (S) in the movable piece is held by the hook when in position.

FIG. 249.—BRASS CATCH TO RETAIN SHUTTER WHEN SET READY FOR USE.

When the shutter is wanted for action, the elastic band, supported by the side loops of wire, is pulled down over the screw (S, Fig. 248) in the centre of the movable piece, which is thus forcibly pulled upwards. It is retained, however, by the brass hook,

H, as shown in Fig. 248 until the right moment, when the bottom of the brass piece (L) is moved to the left, which liberates it, and it is instantly thrown upwards, exposing the lens ; but, on reaching the top elastic band, the screw catches in it and forces the movable piece down again, thus covering the lens.

In fitting both these shutters to the hood of the lens, take the greatest care to cut the hole circular, so that no light may get round the edge. Glue a piece of velvet round the interior of the hole, and, as the thicker and softer the substance into which the lens is pushed the more likely it is to fit closely, glue a piece of sheet cork, 2½in. square, to the bottom half of the ground piece, and cut to fit the lens, which, with the woodwork, forms an additional safeguard.

A NEW DARK-SLIDE.

IN the first place, wood is dispensed with, and vulcanite and cardboard are substituted. The former is obtainable in sheets about a yard square, and the thinnest, which is usually polished, will suffice, though the next thickness, unpolished, is preferable, but of course heavier and more expensive. A sheet of the former costs 3s. 6d., and of the latter 6s., or thereabouts, and both are obtainable from india-rubber warehouses, of which there are several in Cannon Street. Being rolled, the vulcanite is of uniform thickness, and very accurate work can be done with it.

The cardboard must be of uniform thickness also, and about double that of one of the dry plates used ; it may be of the coarsest description, as it will be blackened when used up in the darkslide. An excellent quality is sold by the Willesden Paper Company, Cannon Street.

Measure the camera, so as to ascertain what the extreme external dimensions of the dark-slide are to be. The slide being of a special construction, the back of the camera, into which it is to be fitted, must be adapted to it. The slide must lie against the back, and not slide into it, and must be kept in its place by two buttons, one on each side. The back of the camera must project partly around the slide, and therefore the latter may be described as fitting into the back. The diagram (Fig. 251) shows the form of the back, and the buttons referred to, which retain the slide in its place.

The back of the camera (B) must project ⅜in. beyond that part of it which is marked A. Cover the latter recessed surface with velvet shown dotted on Fig. 251 ; and against this velvet surface the vulcanite slide must rest, and be kept firmly in position, as before mentioned, by two wooden buttons (C C).

The projecting portion of the back, at the top of the camera (B×) must be carefully cut away to a sufficient depth, so that when covered with the velvet the shutter of the slide when open will lie close to the velvet.

A piece of the cardboard specified must be cut the full size of the opening of the camera-back into which the slide is to be afterwards placed (*see* G, Fig. 250, and No. 1 on section). Two pieces must be cut the same external dimensions as the other, but for the centre cut out the exact size of the plates intended to be used (*see* H, Fig. 250, and No. 2 on section), say,

sions, must now be laid on each side of the slide, over the cardboard frame (*see* 3 on section); it may be made of strips of the material, but it is difficult to keep so many pieces in position whilst they are being finally all riveted together. These vulcanite frames must be temporarily stuck in their places with thin glue, to which a little sugar has been added to prevent them from drying quite hard.

Two shutters must be cut out of the vulcanite, the full external dimensions of the slide (*see* 4 on section).

The shutter which has been cut out of the vulcanite, the full outside dimen-

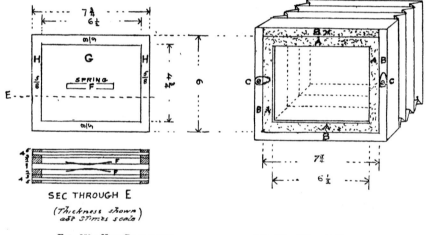

FIG. 250.—NEW DARK-SLIDE.

FIG. 251.—BACK OF CAMERA.

half-plate, or 6½in. by 4¾in., as shown. These latter cardboard frames may be made of strips of the material, to economise it, if desired; but in any case they must be lightly glued one on each side of the first-named piece. The plate will ultimately fit into these framed pieces, so that it is evident the cardboard must be more than the thickness of the thickest plate; it ought to be quite double the thickness, and if the cardboard is not stout enough, two thicknesses of it must be fixed instead of one.

A frame cut out of the sheet of vulcanite, accurately, of the same dimen-

sions of the dark-slide, must have a strip ⅜in. wide cut off the bottom and two sides, but leaving a shoulder ⅛in. wider at the two upper ends. This strip must be cut off in one piece, and may be termed the distance-piece, as it will preserve a space in which the remaining portion, constituting the shutter, will slide up and down. The shutter must be further reduced at the sides to the extent of ⅛in., so as to leave shoulders at the bottom edges corresponding to those at the top of the distance-pieces, (*see* Fig. 252), in which the portion of the shutter to be cut away is etched. The shoulders are intended to prevent the shutter from being entirely withdrawn. The distance-piece must be temporarily glued into the position indicated on plan (*see* No. 4 on section, Fig. '250). Referring to Fig. 252, the distance-piece is marked

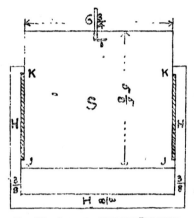

FIG. 252.—SHUTTER SHOWING DISTANCE-PIECE AND SHOULDERS.

HH, the shoulders are marked JJ, the shoulders of the shutter are marked KK, and the shutter itself S.

The shutter only withdraws until the shoulders meet, and it is not hinged. This is not a disadvantage, as the folding-down of a shutter only occupies time, and serves no possible purpose excepting that of appearance; and it is a positive advantage to dispense with hinges, as a source of light-leakage is thereby obviated. Such shutters require, of course, to be gently withdrawn, for fear of knocking off the shoulders, and the thicker vulcanite first mentioned is preferable for the shutter; but as vulcanite is so smooth-running and does not swell and jam with damp, there should be no excuse for breaking off the shoulders.

The shutter must have a small leather tag (T, Fig. 252) riveted in the centre of the top edge, to pull it out with.

The cover-frame of vulcanite (as shown at No. 5, Fig. 250) must be cut identical with No. 3, but without the upper rim, either in one piece or in

strips, and glued to the slide, a slip of stout paper must also be glued over the distance-piece HH to let the shutter run freely. The only object of the gluing is to keep all the various pieces in place whilst they are being bodily riveted together. Before this final riveting is done, the catches and stops must be affixed as follows: The two lower corners of the frame No. 3 (*see* section E, Fig. 250, and LL, Fig. 253) must project as stops, under which the sensitive plate is placed. The plate is further kept back at each side near the top by a small sliding brass catch (MM, Fig. 253), which pushes outwards beyond the dark-slide, and is pushed in again when the plate is in place, so as to catch the edge of the plate about $\frac{1}{8}$in. The spring F (*see* section E, Fig. 250, and Fig. 253) presses the plate against these catches and stops, and so keeps it in register with the focussing-plane. The brass catches may be $\frac{1}{4}$in. wide and a trifle thinner than the vulcanite, and must be embedded into the cardboard frame under No. 3 vulcanite framing. The springs must consist of $\frac{1}{4}$in. crinoline steel about $3\frac{1}{2}$in. long, and, after being bent by hand to a slight upward curvature, they must be inserted through slits in the cardboard.

Small triangular pieces of cardboard, of the exact thickness of the

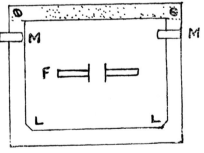

FIG. 253.—FRAME

brass catches, must be glued on under the corners LL (Fig. 253); or still better, the cardboard framing No. 2 may be cut with projecting corners like the vulcanite above it, and then thinned as described.

The slide is now complete all but the upper strip of the cover-frame. This strip, requiring to be very stiff, so as not to bend outwards and allow light to get in when the shutter is withdrawn, must be made of any hard wood, such as walnut. The strip must be ⅝in. wide and ⅛in. thick, and screwed on at both corners of the dark-slide after being sprung to a slight curve by hand, which it will retain, and so always press against the shutter and keep the slide light-tight. Its position is shown dotted on Fig. 253, to which, however, it does not, of course, belong.

All the parts having been accurately glued in position, a number of short lengths of copper wire, as thick as a stout knitting-needle and a trifle longer than the total thickness of the dark slide, must be cut for the rivets. Holes must be drilled through the framing of the slide at proper intervals —say four or five to each side—with a drill just a shade thicker than the wire rivets, and the holes must be slightly countersunk. A rivet being inserted and filed flat at the end, a head must be gently hammered on to it ; then turning the slide over and filing down the other end, a similar rivet-head must be made, but the rivets must not be perfectly clenched until they are all in. The vulcanite will stand a great amount of hammering, but the limit of its endurance might be advantageously ascertained by previously practising on a few spare pieces.

For fixing the portions together, black japan may be used instead of glue ; but whichever is applied, it should be just before the riveting, so that it may be squeezed out by the hammering.

The wooden strips mentioned might be riveted also, but short screws are the best, as the wood may lose its spring and require to be removed and re-bent. The leather tags must have a slot cut out of the framing for them to close into.

The finished slide may now be rasped or planed smooth all round the outer edges, and receive a coat of japan black.

The best way to cut the vulcanite is with a fret-saw, but a straight-edge and steel-edged scraper of any description may be used, and, when nearly through, the material may be broken off.

The finished dark-slide must project a trifle beyond the projecting edges BB (Fig. 251), so that the buttons CC, which must be cut slanting underneath, may jam the slide firmly into place. The surfaces for velvet must be covered with thin glue, and when set, but still tacky, the velvet must be pressed on without stretching it. Velvet against vulcanite forms a perfectly light-tight union, and friction is reduced to a minimum. For large slides, velvet strips may be fixed so that the shutter may be drawn smoothly between them.

AN IMPROVED CLOTHES-HORSE AND CLOTH-RACK.

THIS is very easy to make and the cost nominal, while it can be made any size according to the requirements. The wood used is ordinary deal, the four uprights being of ¾in. wood (1½in. wide and 4ft. 6in. long), with the edges slightly chamfered. In each of these drill four holes of ½in. diameter, the first 1½in. from the top, the other three at equal distances. The rails are cut from the same wood, 1in. wide and 3ft. 6in. long, the edges being well chamfered and sand-papered down so as to round off all corners. The ends of each rail must be reduced and rounded to fit into the uprights. To do this cut a shoulder 1½in. from each end, take off as much wood as you can with a chisel, and then rasp the peg to the right size, allowing it to fit tightly (*see* Fig. 255). These rails are then

FIG. 254.—CLOTHES-HORSE.

3ft. 3in. when fitted. It will be seen that the uprights overlap each other when folded, and the pins of the top rails should therefore be about 1in. longer than the others, as they must protrude ¾in. when the horse is fixed. Taking one side (Fig. 254), it will be noticed that the upright is on the outside at one end, and on the inside at the other, so that they are comparatively locked like a butt hinge, the top rail forming the pin, which must be fixed first. Having fixed all the rails, a single wire nail, with the head cut off, and driven through each pin will hold them in position ; punch the nails well in, and then saw off the protruding

FIG. 255.—END OF RAIL.

pins of the second and bottom rail flush, and clean off with the smoothing-plane. The pins of the third rail are utilised for holding the cross-pieces which regulate the opening of the horse. These are more clearly shown in Fig. 256, and measure 21in. long, 1½in. wide, and ⅝in. thick, with six

or more notches in each. They must fit tightly on to the pins, and will not require any other fixing. For the pins of the top rail, which protrude also, procure two hardwood knobs by way of finish, and fixed them firmly on. Glue must be entirely avoided, and nails must not appear anywhere, as they will iron-mould any clothes which are damp. When not in use this novel clothes-horse folds up nearly flat, and therefore takes up very little room. If the horse is required on a larger scale, the wood should be proportionately stronger, the uprights being of 1in. stuff. Smaller sizes than the one given are very useful.

Fig. 257 shows a very handy rack for permanently fixing to the kitchen wall on which to hang tea and other cloths for domestic use. Two battens about 2ft. 6in. long, 3in. wide, and ¾in. thick,

FIG. 256.—RACK.

are fixed to the wall at any distance apart, according to the space available. The drawing shows the rails 2ft. 9in.

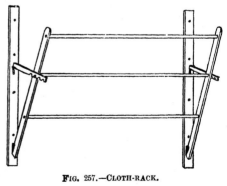

FIG. 257.—CLOTH-RACK.

long, which is a useful size. The uprights to which the rails are fixed are 2ft. long, 1¼in. wide, of ¾in. stuff, the rails being fixed in the same way as explained for the clothes-horse. The bottom ends of the uprights are hinged to the battens with brass strap hinges, so that should occasion require, the rails can be folded up close to the battens and secured by a small brass hook. The side-pieces for regulating the angle of the rails are also hinged, and the length of them depends upon how far the rails are required to project. The ends of the middle rail are left projecting, and form the stops for the racks.

A HANGING WHATNOT.

FIG. 258 represents a whatnot with an occasional writing-flap, for which purpose it is well suited ; but it will perhaps be found better adapted for a receptacle for music, periodicals, or loose engravings, etc. In the latter case it will be more easily made than if applied to the former use, as the cut in fretwork, a suitable design for which, in the form of an oblong panel, there would be no difficulty in obtaining.

Fig. 259 represents the back, or foundation of the whole structure. It is a shield-shaped board, either of oak, American walnut, or any other wood usually employed in cabinet-making.

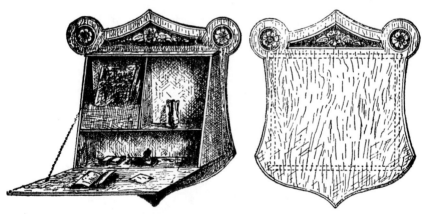

FIG. 258.—WHATNOT WITH WRITING-FLAP.

FIG. 259.—BACKBOARD.

internal arrangement will be much simpler, consisting of a thin board running across the opening and reaching half the height—*i.e.*, 9in. This is meant to keep the music, etc., in position, and at the same time to allow of easy access to the contents. Should this form be adopted, the board mentioned might be ornamented by being

The thickness of the back must be at least ⅜in. finished work, and whatever jointings are necessary to make up the required width must be accurately made and carefully secured by dowels and glue. The shape of the shield having been carefully drawn, cut out as sharply as possible with a suitable saw, finish up the outline with a spokeshave, sand-paper, and chamfer all round as in Fig. 259. Finish the surface with a smoothing-plane and fine sand-paper, and then set out the positions of the various parts and ornamentation as shown by dotted lines in Fig. 259. The

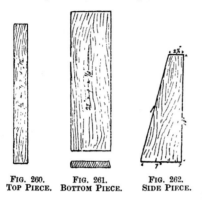

FIG. 260.   FIG. 261.   FIG. 262.
TOP PIECE. BOTTOM PIECE. SIDE PIECE.

ornamentation should be carved if the maker possesses the necessary skill; but if not, or if other decoration is preferred, some simple designs worked in repoussé brass or embossed leather are most effective.

The top, bottom, and two side pieces (Figs. 260, 261, and 262), call for little description, as the shapes and dimensions are evident from the illustrations. All these pieces must be secured to the back and to each other at their several

FIG. 263.—FALLING-BOARD.

points of juncture, by being "let in" to shallow rebates, say ¼in., and gluing. They must also have several 2in. screws driven from the back of the shield—say three screws to each of the pieces. The bottom piece must

also be fixed to the sides by similar screws, two to each side, driven, of course, from underneath. The top piece being so narrow, and not being subject to any great strain, will be sufficiently secured by the means already described.

The folding front (Fig. 263) is composed of a board 23in. by 20½in. This

FIG. 264.—SECTION OF FALLING-BOARD, SHOWING FRAMING.

board must be of ⅝in. stuff, the top and two side edges being finished with a simple double beading as shown in the section Fig. 264. In order to provide against warping, and also to give finish to the falling-board, a false framing of stuff about $\frac{7}{8}$in. thick, and 2in. or 2½in. wide, and rounded on the outer edges, should be placed around it, as represented at Fig. 263, a section of which is also shown at Fig. 264. The inside edges should be chamfered as also indicated. This framing may be secured by gluing, and allowed to set under pressure.

The falling-board may be ornamented by a floral or other design painted in

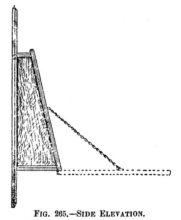

FIG. 265.—SIDE ELEVATION.

oils, by leather work, repoussé work, etc., according to taste, or even by a suitable piece of Lincrusta Walton.

This also applies to the ornamentation of the upper portion of the shield, as this material lends itself to such uses admirably.·

The folding board must be fixed in position by a pair of suitable brass butt hinges, and provided with brass chains to support it as represented in Fig. 258, and in the side elevation, Fig. 265.

A THREE-CORNERED WRITING-TABLE.

FIG. 266 shows the writing-table when finished. Its object is to fill a corner of a room without occupying much space, for which reason the top of the table is fitted with a flap, as shown by the dotted lines in Fig. 266, by raising which a sufficient

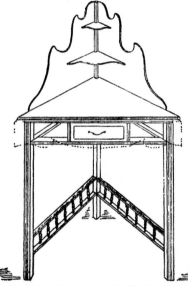

FIG. 266.—THREE-CORNERED WRITING-TABLE.

area for writing on is obtained, whilst at the same time the table only measures about 15in. from back to front. The sides above the top, to which the small shelves are fixed, are for both use and ornament. The top shelf will hold a vase or fancy pot, the lower one the inkstand.

First cut the top out of ⅝in. yellow pine, 2ft. 6in. long on the front, with a depth of about 15½in., which will

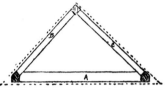

FIG. 267.—RAILS WITH SECTIONS OF LEGS.

bring the remaining two sides at right angles to each other. The legs are 2ft. 4in. long, the back one being 1in. square, the front legs 1in. by 1¼in., on account of their peculiar shape. The necessity of their being so will be seen by referring to Fig. 267, giving sections of the legs. The front rail, A (Fig. 267), is 1½in. wide, of ¾in. stuff, and measures 2ft. long when fitted, the ends being mortised into the legs ½in. each end, the front edge being flush with the front of the legs; the side rails, B and C, are 1in. wide, ¾in. thick, and 17in. long when fitted, being mortised into

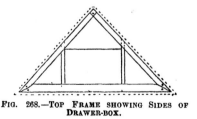

FIG. 268.—TOP FRAME SHOWING SIDES OF DRAWER-BOX.

the front and back legs in like manner, these rails being 3in. from the top of the legs. Fig. 268 is a similar frame, only it is screwed *on* to the tops of the legs, the corners being mitred on the front legs, the side rails overlapping each other on the back leg by what is known as "halving," through which a screw is put; but nothing must be

FIG. 269.—FALSE SLIP.

fixed permanently until the fancy rails at the bottom are made and fitted.

The space allowed for the drawer is 10in., and the sides of the drawer-box must extend to the rails at the back, and be fixed by screws through the top and bottom front rails and also between the side rails. This will be more readily understood by referring to Fig. 268. The grain of these sides must run horizontally for strength; at the same time the grain should be vertical on the front edge, and in order to do this a false slip must be glued on

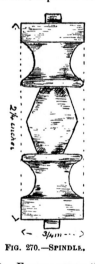

FIG. 270.—SPINDLE.

(Fig. 269). For one reason the whole of the front frame and legs are ribbed or beaded, and it is easier to cut a bead with the grain than on the cross.

The fancy rails at the bottom must next be made, each side consisting of

two rails, 17in. long when fitted, 1in. wide, ⅝in. thick, and seven spindles. The top rail is beaded on the top and front edge, the lower rail on the front edge only, allowing ½in. each end of the rail for mortising. The spindles (which should be turned out of hard wood such as beech) are 2¼in. long when fitted, allowing $\frac{3}{16}$in. each end for fitting in the rails, and are ¾in. in diameter (*see* Fig. 270). The spindles are fixed 2½in. apart from centre to centre, and glued to the rails.

Next proceed to fix the frame together. See that the mortises fit well,

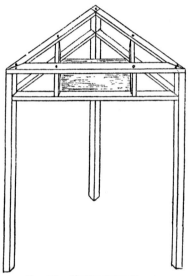

FIG. 271.—FRAME WHEN FITTED.

and then glue will be sufficient to hold the frame firm, particularly as the top part is screwed on to the top of the legs. The fancy rails should be 6in. from the bottom. Fig. 271 shows the general appearance of the frame when fitted; a thin board being screwed under the sides of the drawer-box for the drawer to slide on. The drawer measures 10in. square and 3in. deep (outside measurements), the front being ¾in. thick. A good plan to take off the flatness is to cut a rebate on the outside, say ⅝in. wide and ⅛in. deep. Use a rebating-plane with the grain, and a thin tenon-saw and chisel for the cross-grain. A fancy brass drop handle should be provided ; also a stop glued at the back for the drawer to shut against.

It is hardly necessary to describe how to make the drawer. The fancy

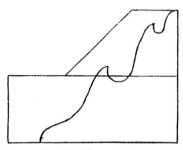

FIG. 272.—FANCY BACK.

back must be made of thin wood, say ⅜in. The extreme height of the back is 18in., taking two 9in. boards, each side being 19in. long. The easiest way is to cut three boards, say 2ft. long, and having planed them up, cut one in halves diagonally, as shown in Fig. 272, and join half to each of the remaining boards. Having done this, proceed to mark out the shape in pencil, or, better still, cut a paper or cardboard pattern of one side, and mark the wood from that—this will ensure both sides being alike — then cut it out with either a frame saw or fretsaw. The sides constituting the back are screwed to the table-top at the edges, and one must overlap the other at the back. To strengthen them, fit an upright ¾in. square, beaded in character with the other parts of the table, on the two sides showing, and then screw the back to this ; and in the event of the wood warping (as on account of the thinness

it is very likely to do), by screwing the sides to the upright it will draw them perfectly straight, and the table will present a very neat appearance (see Fig. 266).

The two shelves are screwed on from the back, and must also be of ⅜in. wood, the lower one measuring 9in. long on each side, the top one 5in. each side, the fronts being cut out circular or otherwise ; they are fixed 6in. and 12in. from the table-top respectively.

The flap for the front (Fig. 273) is 2ft. 6in. long, 4in. wide, and the same thickness as the top, viz., ⅜in. wood. This is screwed to the front of the fixed top with a pair of 2in. polished brass hinges. The hinges are let into the edge of the flap, so that when fixed up for use the flap and fixed top may appear as one board ; and this is easily managed with a little care. The best way to fix the permanent top is to lay it on the bench upside down, and place the framework on it, and screw through the upper rails, taking care that the points of the screws do not come through. The top must overlap the framework ⅜in. in front, and about ½in. at the sides. The extra depth in the front is to allow for two wood brackets screwed on to the front of the legs to support the flap when in use. The brackets must fold back against the front of the frame when the flap is

FIG. 273.—SCREWED FLAP.

down and be quite hidden, so the size must be judged accordingly.

With regard to the finish, it looks very well sized and varnished (best copal varnish), or it can be enamelled in any Art colour.

A FANCY SHELF-BRACKET.

THE illustration (Fig. 274) shows a simple and yet effective pattern for a shelf-bracket, most suitable for enamelling and of easy construction.

The top is 9in. long, 6½in. wide, of ½in. wood, the front corners being cut off as shown, and the edges marked

FIG. 274.—FANCY SHELF-BRACKET.

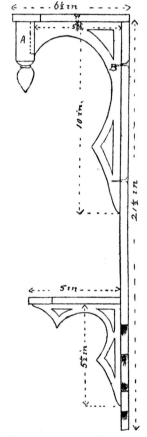

FIG. 275.—SECTIONAL SIDE VIEW OF SHELF-BRACKET.

It looks well in ivory-white for a drawing-room, and equally as well in black-and-gold.

with a bead-router. The back is of one piece the same thickness, 21½in. long, and 7in. wide; the space for the glass being cut out 8in. by 4½in., 5in. from the top. A thin piece of wood should cover this opening by screwing it on from the back, the glass being fixed in from the front and held in position by a thin beaded moulding ½in. wide, fixed on the front (the corners being mitred) with glue and fine brads or needle-points. To increase the effect the glass should have bevelled edges.

Before fixing the top on to the back refer to Fig. 275, a sectional side view of the shelf-bracket. The end pieces of the cornice marked A are 1in. square and 2in. long, the front and outer sides being fluted by means of a fine gouge; the crossbar is ½in. thick, ¾in. wide, and must be let into the end pieces and glued, leaving a space of 1¼in. between it and the top, which will be the length of the five spindles minus the small pegs at each end to let into the crossbar and the top. These spindles are fixed 1in. apart, measuring from centre to centre, and should be turned out of hard wood, such as beech, ½in. in diameter. The pegs on each end of the cornice are ¾in. in diameter and 1½in. long; and corresponding holes must be drilled into the crossbar and the top, at the distances mentioned, to receive them.

The back of the end piece A must have a groove ¼in. deep and ⅜in. wide, into which the brackets B are let (*see* Fig. 275). The cornice can then be fixed by gluing the pieces A to the top, and also by a screw in each from the top. The top is fixed to the back by two screws. The cornice is fixed ¼in. from the front. The two brackets B are 10in. by 5¼in., exclusive of letting into the front piece A; they are cut out of ⅜in. wood, and fixed by two screws from the back and one from the top

to each, and can be much improved by using the bead-router on the front edges similar to the top and the shelf.

The shelf is 7in. long and 5in. wide, the front corners being cut off to correspond with the top and fixed to the back by two screws. The bracket or its support is of similar pattern to the brackets B, and measures 5½in. by 4¾in. ; the top part must only be glued to the under part of the shelf, as a screw-head would show, but secured to the back with two screws. All screws must be as thin as possible, and 1in. in length. The whole work can be made of ordinary deal, with the exception of the spindles and knobs, so that the heads will screw in flush without countersinking and not be visible.

Be careful to sand-paper each part well before fixing together, and to paint over any knots in the wood with a preparation known as " knotting." If required to be in black-and-gold, French polish and vegetable-black mixed to the consistency of thin paste make a good black, of which there should be two coats. The first coat, when dry, should be rubbed down with the finest sand-paper to obtain a perfectly smooth surface. After having applied the second coat, decorate the parts required with Judson's gold paint. This will impart a good appearance, in dead black-and-gold. If it is desired to have a polished surface, varnish the work with hard crystal or best copal varnish. If enamel is preferred, it is advisable after applying the knotting to give a first coat, consisting of white lead, turps, and a little gold size, to harden it. This forms a ground-work ; but it should be rubbed down with sand-paper before putting on the enamel.

A SIDEBOARD.

IG. 276 may at the first glance seem formidable for an amateur to attempt ; but if he will take the trouble

The cupboards (Figs. 277 and 278) must first be made. They measure 2ft. 6in. high, 18in. wide, and 18in. deep

FIG. 276.—SIDEBOARD.

to follow the description of how to make it, he will see how much easier it is to make than it first appears.

(all outside measurements). It will be seen that these stand independently on the right and left hand, and when placed in position have a space of 18in. between them, thus making the total

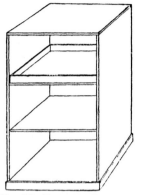

FIG. 277.—LEFT-HAND CUPBOARD.

length from outside to outside 4ft. 6in. The cupboards are made of ⅝in. wood,

shut on to the face of the cupboards, so that a plinth the thickness of the door must be screwed on to the three sides of each cupboard (from the inside), the corners being mitred and a ¼in. bead run on the top edge. The backs are of common wood stained (⅜in. thick), and are screwed on. The left-hand cupboard (Fig. 277) is fitted with a shelf and a sliding tray for glasses, the tray running on fillets, the shelf being a fixture. The right-hand cupboard (Fig. 278) has a centre shelf only, and a drawer or cellarette for wines; this must be made ½in. less in depth than the cupboard, to allow for the knob. The drawer should have a ⅝in. or ¾in. front, with ½in. sides and back, and be fitted with divisions for bottles or decanters, or it could be left plain.

The next part to be considered is the frame for the three drawers: this will rest on the top of the cupboards, the sides, front, and back being flush with the same. The measurements

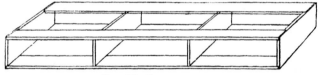

FIG. 279.—DRAWER-FRAME.

the tops and sides being dovetailed together. The bottoms, which are

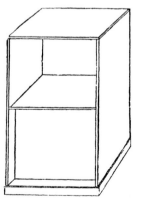

FIG. 278.—RIGHT-HAND CUPBOARD.

2½in. from the floor, rest upon fillets of wood screwed to the side. The doors

are: length 4ft. 6in., height 6in., depth 18in. With the exception of the sides the frame is of common deal, as it will be entirely hidden. Fig. 279 will give the reader an idea how to make it. The sides must be of American walnut, ⅝in. thick., or whatever wood has been decided upon. The remaining

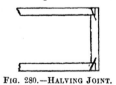

FIG. 280.—HALVING JOINT.

parts are of deal; the rails being of ¾in. stuff and about 4½in. wide, and the divisions ½in. thick. The thickness of the sides will allow of a rebate at the top and bottom, the thickness of the rails, which afford a slight shoulder against which the top and bottom boards can be fixed (Fig. 280).

M

The partitions are of course 6in. high, being cut out at each end for the rails to fit on to (Fig. 281), and are so fixed that the centre division is 18in. ; a back is then screwed on similar to the cupboards. This frame will rest on the top of the cupboards ; it must not be made a permanent fixture, but blocks, the thickness of the rails, must be

FIG. 281.—PARTITION OF DRAWER-FRAME.

screwed on the top of the cupboards in such positions so as to prevent any slipping.

The drawers are made in common wood, fronts of ¾in., sides and back ½in., and must fit flush with the front of the framework. The drawers must have false fronts to shut on to the edges of the frame, and of the same substance as the doors, with which they must fit flush ; these fronts are 6in. by 18in. and ¾in. thick, and can be easily fixed to the drawers themselves with screws from the inside. The end drawers will cover the edges of sides, rails, and partitions, and the middle drawer will conceal the top and bottom rails, so that when shut the whole of the front of the frame is hidden.

The next parts to be considered are the doors. These are of the simplest kind, the uprights and rails being 4½in. wide by ¾in. thick, allowing for the depth of the plinths (2½in.) ; the doors measure 2ft. 3½in. by 18in. Fig. 282 shows the position of the tenons, which, it will be noticed, do not appear through the uprights, but must be at least 2½in. long. The inner edge of the frame is grooved to admit of a panel about ¾in. thick, which should be of one piece if possible. The edges on the front are also chamfered, to take off the plainness.

Brass butts, 2in. or 2½in. long, are used for hanging the doors ; and it is hardly necessary to explain how to fix them. The doors, when shut, will be flush with the drawers.

The most important part is the top-board. No matter what wood is selected for it, it will require most careful

making. The length is 4ft. 9in., which allows 1½in. overlap at each end, the width 1ft. 9in., and it should be 1in. thick if possible. The width will necessitate one or more joints. For these, the edges to be joined must be shot perfectly true and glued ; and for additional strength, strips of wood ¾in. thick, 4in. or 5in. wide, and of such length as to fit in between the rails of the drawer-frames, must be screwed on underneath. The boards must be well clamped together until the joints are thoroughly firm. When the clamps are removed, finish off the top with the smoothing-plane ; and having squared the edges, a moulding should be run round, the pattern of which is left to the amateur. The one drawn in Fig. 276 is the commonest.

The topboard is a fixture ; that is, it is screwed to the drawer-frame ; but, before this is done, the supports for the shelf must be screwed to it. These columns are 12in. high and about 3in. in diameter at the largest part. There are several patterns to select from, at prices ranging from 3s. 9d. each upwards. The shelf measures 4ft. 6in. by 10in. and is of ⅝in. or ¾in. stuff, with a moulding to

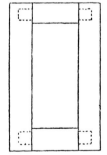

FIG. 282 —PLAN OF DOORS, SHOWING POSITION OF TENONS.

correspond with the topboard. The back of the sideboard—between the topboard and the shelf—is a frame made up similar to a door running the length of the board and 12in. high. The rails are 2in. wide, end uprights 5in. wide, and centre upright 6in. wide, leaving two spaces 18in. by 8in. for mirrors, so that the backs must be rebated sufficiently to take the glass, the front edges being chamfered like the doors. The rosette in the centre is 4in. in diameter, and costs from 8d. to 1s. ; the tenons, as in the doors, do not appear through the rails, providing they are long enough to hold well when glued. The mirrors should be of silvered plate-glass, with bevelled edges, which are easily obtained cut to size, and when fitted in their places a back of thin wood is screwed on. The glass back is fixed on the top of the board as far back as possible by three or more hardwood pegs ; the shelf rests on the same, and is held firmly with screws. These are in their turn hidden by the carved top, which is 40in. long, and held to the shelf with wooden pegs, and, for additional strength, a fair-sized strip screwed down the back. To fix the columns to the shelf small dowel screws are fixed in the former, then screwed into the shelf, and also screwed to the topboard from underneath at a slight distance from the ends. The board can now be made secure to the drawer-frame with short stout screws from underneath, taking care the points do not come through.

For the doors four carved uprights, and for the drawers four short ones, are required, the latter being fixed on the *end* drawers. A thin beaded moulding 1¼in. wide is fixed at the bottom of the fronts between the carved work, and it is hardly necessary to say it must be of the same wood as the rest of the sideboard, and held by glue and fine brads or needle-points. The columns for the doors can be bought of the right length, viz., 2ft. 6in. ; they must be cut through at the bottom where the plinth comes, so that when the doors are shut they appear as one piece, and can be screwed to the doors from the inside.

Two brass locks and three pairs of fancy handles will complete the sideboard. The whole may be French-polished, and the amateur should let this be done by an expert polisher, a good finish being most important.

The carved woodwork can be bought at The Builders' Supply Stores, 145, Holborn Bars, London, E.C., the proprietors of which supply an illustrated list of various kinds of woods and patterns.

With regard to the cost. The carving shown in Fig. 276 costs about 30s. in beech, birch, or limewood, and about 45s. in oak, walnut, or mahogany.

The total cost of the sideboard if made in American walnut would be well within £5. In yellow pine, which could be stained in imitation of oak or walnut, the cost would not exceed £3.

A DINNER-WAGGON.

THE illustration (Fig. 283) shows a useful kind of dinner-waggon mounted on three wooden wheels, 6in. in diameter, and 2in. wide, which are grooved on the edge, and round the grooves indiarubber bands are stretched so that the movement shall be silent. Two of the wheels are fixed at one end, one on each leg ; the third wheel is so fixed at the other end that it works loosely on a pivot in order to allow the waggon to be moved about in any direction easily.

The legs are 4ft. long, 2in. square, and must be of hard wood if turned similar to Fig. 283. If the amateur possesses a lathe, and can do his own turning, it is doubtful if the bed of the lathe will be long enough to take a 4ft. leg, and if he is unable to extend the bed each leg must be turned in two pieces. It must be planed up square first, and the parts to be left square must be ribbed or fluted with a hand-router. Fig. 284 will show more readily how each leg is divided out,

M 2

and where it is divided for the convenience of turning, being joined again by screwing in a stout dowel-

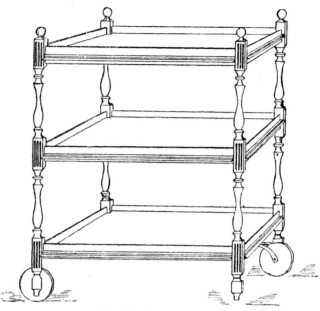

FIG. 283 —DINNER-WAGGON.

screw half in each piece, and gluing in addition.

Should the amateur not possess a

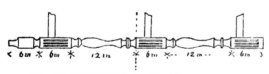

FIG. 284 —DIVIDED LEG FOR TURNING.

lathe and yet be desirous of making a dinner-waggon, Fig. 285 is a suggestion that will answer the purpose, though a turned leg is preferable.

The length between the legs is 3ft., and the width (also between the legs), 18in., so that each of the three shelves measures 3ft. 4in. by 1ft. 10in., outside to outside.

Each shelf rests on rails 1½in. wide by ⅝in. or ¾in. thick, mortised into the

legs on the sides and ends. The shelves themselves are of ⅝in. wood, each shelf having one joint on account of the width. The shelves have a moulding run round the four sides similar in pattern to what is generally found on a dining-room table : and for additional strength each shelf should have two pieces of wood or stretchers, say 3in. wide by 18in. long, screwed from side to side underneath. The rails for the top shelf are mortised in 4in. from the top, the middle rails 1ft. 9in., and the bottom rails 3ft. 3in. from the top,

FIG. 285.—LEG MADE WITHOUT TURNING.

each rail having a ¼in. bead run on the lower edge.

Having made the shelves, a piece 2in. square must be cut out of each corner to fit inside the legs. Each shelf must be fitted carefully in its respective place. Having done this, and before fixing the shelves permanently, a ledge, say 1¼in. deep and ⅜in. thick, is fixed on all four sides of each shelf and between the legs, as shown in Fig. 283, being screwed to the shelf from underneath. The shelves are fixed to the rails with screws through the inside of the rails diagonally. It is essential that the rails should be firmly fixed into the legs, as, in consequence of the waggon being wheeled about, any weak parts would soon be found out, and these must be avoided.

The wheels are so fixed that at any time they can be removed and the waggon rest on its own legs. Most amateurs will probably be able to turn

FIG. 286 —FIXING OF REAR WHEELS.

the wheels themselves : they must be of hard wood, such as ash, beech, or oak, 6in. in diameter and 1¼in. wide, with a groove ¾in. wide and ¼in. deep to hold a rubber ring or band. The two wheels at the one end will work on iron bolts fixed in the legs, the bolts having square shoulders so that they will not turn round, and what are known as cheese heads that will let in flush with the wood. Each bolt is 4in. long and ½in. thick, and in place of the ordinary nut procure a winged or fly-nut in either brass or iron. The hole in the wheel should only be slightly larger than the size of the bolt, so as not to allow too much play, an iron washer being put on each side of the wheel (see Fig. 286).

The third wheel will possibly require the aid of a blacksmith, as it is necessary that it should be fixed on a swivel-joint, in order to more readily guide the waggon when being wheeled. If it had a wheel on each leg, one end would have to be lifted slightly when turning it, or else the strain on the

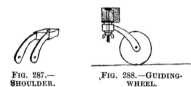

FIG. 287.— SHOULDER. FIG. 288.—GUIDING- WHEEL.

wheels would be great, particularly when the waggon was full ; but a guiding-wheel avoids this altogether. The ironwork for this wheel (Figs. 287 and 288) is similar in shape to the lower part of a socket-caster. The side-rail which supports the guiding-wheel must be strengthened by screwing a piece of wood to it from the inside, 12in. long, 1in. thick, and the same depth as the rail. Through this the bolt is fixed, the head being let in flush ; and it is as well to fix a small iron plate underneath this piece of wood for the frame of the wheel to turn on. Any blacksmith would be able to make it easily, and it is necessary to fix this before fixing the bottom shelf.

The knobs (Fig. 289) on the legs or uprights should be rather larger in diameter than the legs, say 2½in., the base being 2in., and a short plug left on to be glued into the legs.

FIG. 289.—KNOB.

The waggon, if made of ordinary wood and stained to imitate walnut, oak, or mahogany, should be varnished. Avoid spirit varnishes ; there is nothing better than copal for this work. Give it two coats after sizing, rubbing down the first coat before applying the second. If either of the above-mentioned woods are used the waggon should be French-polished.

A COMBINATION TABLE AND MUSIC-STAND.

THIS forms a useful occasional or fancy table (Fig. 290), and when folded up converts itself into a music-stand (Fig. 291). The music-stand can be regulated to any angle to suit the player, and if necessary the whole thing can be made to fold up quite close, occupying very little space. The ledge that holds the music is made to let down on to the legs when the table is in use ; and when the music-stand is finish it off with enamel, then the cheapest wood will do.

The top is 21in. long and 18in. wide, made up of three boards, the outer ones being 9in. boards, which really measure about 8¾in. when the edges are planed. The wood is ⅜in. thick when planed, and this will be found quite strong enough, as the table is not intended for rough use nor for holding heavy things.

FIG. 290.—THE TABLE.

required, the ledge is held in its place by means of a wooden button screwed underneath.

The whole structure is very light, and if carefully made is perfectly strong and rigid, even if ordinary deal wood is used in making it   The actual cost in this case would not exceed a shilling, but such woods as walnut, oak, mahogany, or pitch-pine could be used and polished ; but if the idea is to

Having planed the wood both sides, the boards must be laid face downwards on the bench and four cross-pieces 1in. deep and ¾in. thick screwed on (Fig. 292). The pieces marked A extend over two boards, and are screwed to them ; the pieces marked B are screwed to the remaining board, but not to the centre one, and measure about 11½in. long. It is better, before screwing the two boards to the cross-piece A, to glue the edges and clamp them together, and when dry run the smoothing-plane over them

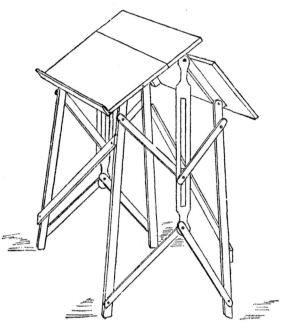

FIG 291 —THE MUSIC-STAND.

to get a true surface. It will be noticed that the pieces B fit close

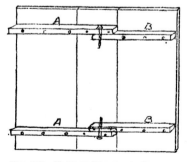

FIG. 292.—PLAN OF TABLE-TOP, SHOWING UNDERNEATH SIDE.

against the pieces A on the inside, the latter being 1¼in. from the sides of the top. In order to allow the table to fold up, the top, which is practically in two parts, must be hinged in the centre, which will be 10½in. from each end, and a swivel-joint formed with screws. The screws are made a fixture in the pieces marked B, the holes in those marked A being large enough to allow the joints to work freely. For these particular joints the screws should be 2in. No. 12, with either round heads or countersunk. If the latter, they must be let in flush with the wood. The former are, however, better, a small iron washer being put between the head and the wood. The centre guides are hung on these screws, of which more will be mentioned later on.

The legs must next be cut out, and are 2ft. 3in. long, 1in. wide, and of ¾in. wood, which practically means about ⅝in. when planed; the top ends must be rounded, and the edges slightly chamfered; 1¼in. screws are used for fixing these. It will be noticed, by referring to Fig. 290, that the legs extend outwards at the bottom when fixed. This ensures a steadier table, and likewise when the frame is contracted to form the music-stand. Should the table be required to fold up closely, these legs must be fixed nearly vertical, and the stays (especially the bottom ones) will be considerably shorter. The centre guides, in which the top stays work, are made up of three pieces, glued together and pinned through. Fig. 293 will give the reader some idea of how they are made. The two outside pieces are ½in. wide, and the middle pieces ⅜in. wide and ⅝in. thick, a piece being fixed at each end, leaving a space of about 17in. for the stays to slide up and down. The ends of each guide are then finished off, as shown by the dark lines in the illustration, and a hole drilled near the top to

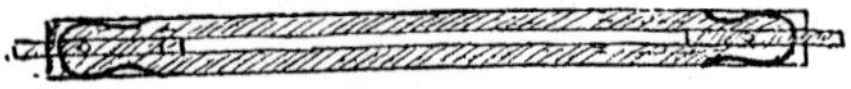

FIG. 293.—STAY-GUIDE.

fix them to the cross-pieces of the table-top, similar holes being made at the other end for the bottom stays.

If possible, all the screw-holes should be drilled, as the wood is thin and narrow, and, consequently, liable to split. The amateur must also be careful that each leg and stay is drilled exactly alike. The stays are of ⅜in. wood, 1in. wide, the top ones measuring 11in. long, the bottom ones 13½in. ; these must be cut out and the screw-holes drilled ready for fixing, when the parts may be put together.

The legs are next fixed on the cross-pieces 1½in. from the ends. Those on the cross-pieces A are fixed on the inside of them, and the others on the cross-pieces B on the outside. Next fix the

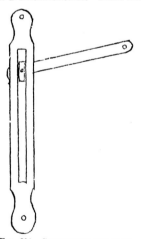

FIG. 294 —STAY-GUIDE AND STAY.

stay-guides, which are screwed on the outside where the top is hinged ; 2in. screws are used here on account of the three thicknesses of wood.

Attention must now be turned to the stays, for upon the proper fixing of these depends the true working of the whole frame. The ends are screwed to the legs on the outside ; but the other ends that are fixed to the guides are fixed one back and one front, with the guide between them, which is clearly shown in Figs. 290 and 291.

Now, the bottom stays are, as it were, fixtures, whereas the top ones slide up or down in the slot formed in the guide ; a small piece of hard wood, about 1in. long, to which the stays are attached by means of a screw, works in the slot (see Fig. 294, showing the guide and one stay). A stretcher-bar at each end, 1in. wide, ½in. or ⅝in. thick, fixed at an angle, will keep the legs firm, and is simply screwed on to the legs.

The music-holder is a simple contrivance, made of the ⅜in. wood, 1¼in. wide, extending the width of the table.

FIG. 295.—MUSIC-LEDGE.

This is fixed to the cross-pieces A on the outside with two small pieces of wood about 4in. long, so that when the table is used, the ledge appears as shown in Fig. 290 ; and by fixing a small button either of wood or metal underneath the table, the ledge is kept in position when required for a music-stand.

All edges should be slightly chamfered, and for an inexpensive stand the following is a good finish in black and gold. First clean the wood well with fine sand-paper, then make a mixture of vegetable black with water and a little size, and apply this evenly, doing all the edges first and the surface last. When this is dry and hard, rub it over with sand-paper and then give it a second coat. Relieve the chamfered edges with gold paint.

Finally, French-polish, or, if this is beyond the amateur's power, a coat of spirit varnish will look nearly as well. If after the second coat of black it is found the work is patchy, it must be rubbed down and done again, otherwise the varnish or polish will show up the defects.

A FOLDING JEWEL-CASE.

THIS jewel-case is made in the form of a small chest of drawers; but instead of the drawers sliding out, the four top ones are hinged at the back and the outside finished very smooth; a small quarter-round or other moulding is fixed round the top at the front and the two ends, projecting slightly as shown.

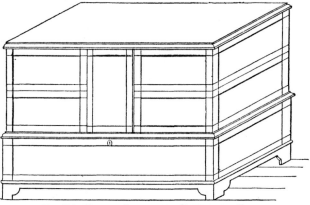

FIG. 296.—FOLDING JEWEL-CASE—PERSPECTIVE VIEW.

corners, and open outwards. The lower one is intended for larger articles, and is not divided: thus there are no trays to lift out, and any compartment can be opened by simply raising the lid and pushing the required compartment to one side.

Fig. 296 is a perspective view, and Fig. 297 an end section showing the lid open; Fig. 298 is a plan view showing two of the compartments open. The length is 9in., the width 4½in., and the height 7in.

It is made of mahogany, stained and polished a dead black, and the grooves or lines on the outside gilded. The lower compartment is made of ½in. wood and dovetailed; the back is carried up to the height of the two top compartments, as shown in Fig. 297; the bottom is ¼in. thick, and has a bead worked on the front and two ends, which projects as shown. The dovetailing must be neatly done,

The four top compartments are made of ⅜in. wood and dovetailed. The out-

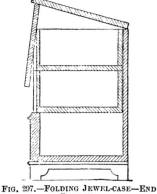

FIG. 297.—FOLDING JEWEL-CASE—END SECTION.

sides must finish ⅛in. within the moulding round the bottom compartment, and ¼in. clear of each other in the centre; they must be made exactly of one size. A recess ⅛in. deep is cut in the front to admit the piece hinged to the lid; this recess is shown in Figs. 297 and 298. The bottoms of the compartments may be of pine, fixed in a groove or rebate worked in the sides. These four compartments are fixed to the back by brass hinges the full height of each and let in flush, both into the back and the compartments.

The lid is made of ⅜in. wood, with a quarter-round moulding, and worked

and the lock is fixed in the centre of the bottom compartment. When locked, this piece of wood will secure the top compartments. The top arris of the recess is slightly cut away to allow this piece of wood to work more freely. A brass escutcheon is fixed over the key-hole.

The grooves in the front and ends are made with a router; they are ₁/₁₆in. wide and ₁/₈in. deep. In cutting across the grain, a piece of wood is cramped against the outer edges to protect them from being damaged.

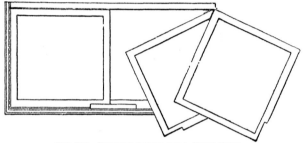

FIG. 298.—FOLDING JEWEL-CASE—PLAN VIEW.

round the edges. In working the mould across the ends a piece of wood is cramped against the outer edges, to prevent them from being damaged; the lid should project ⅛in. at the front and the two ends. The lid is fixed to the back by two brass hinges, which must be let in flush. A piece of wood ¼in. thick is fixed to the under-side of the lid by a brass hinge, as shown in Fig. 297; it must be made the width of the recess in the front of the top compartments. The lock-staple is fixed to the lower end,

The feet are cut out to the form shown, mitred at the corners, and fixed by glue and screws from the under-side. The outside must be smoothed over with glass-paper stretched on a piece of wood, leaving the angles and arrises perfectly square. The outside is stained and polished a dead black, and the grooves are neatly gilded. The insides of the compartments are lined with velvet: cut the velvet to the exact size, brush the inside over with thin, hot glue, and carefully fix the velvet in its place.

AN OLD-STYLE CABINET OR SIDEBOARD.

FIG. 299 represents the front and Fig. 300 the end elevation of an old-fashioned cabinet or sideboard such as is still to be met with in many country places.

Most of the details given are true to an original old Welsh cabinet.

The extreme measurements are 4ft. 6in. wide, 1ft. 9in. back to front, and 6ft. 8in. high.

The bottom, being the principal part, must receive the first attention, merely premising that before any portion is actually begun, a drawing of the whole will be set out by the maker.

This portion is 4ft. 2in. wide, 1ft. 8in. back to front, and 3ft. high over all. The top is full 1in. thick, is moulded as in Fig. 301, and overhangs ¾in. in front and at the ends. As will be seen from Fig. 300, it does not let the back which comes under the middle portion be of pine with oak ends clamped or otherwise fastened on. The corner-pieces are 3in. wide in front and 1¾in. at the ends. Those at the back are 2in. square. In order to save material, the front and ends may be framed up in the ordinary way, and if necessary a bead be run to break the joint where they are fastened together. The rail between the

FIG. 299.—FRONT ELEVATION OF CABINET.

FIG. 300 —END ELEVATION OF CABINET.

extend further than the front of the middle portion which fits on behind it, but it will be better to carry the top right to the back, and so hide the rather ugly joint, which is otherwise too conspicuous. The top need not be carried back in a solid piece, as it will be sufficient for it to be wide enough to go under the front of the middle portion, and have pieces a few inches wide continued at the ends to the back ; or, if an entire top be preferred, drawers and the doors is 1⅝in. wide exclusive of the mouldings above and below it. The bottom rail is of the same width with the piece which Figs. 299 and 300 show projecting from it. The upright pieces between the doors and drawers are 2½in. wide on the flat. Above the drawers it will be noticed that there is no bearer. They work directly against the top, which is just blocked on, but it will be better, especially if the top runs to the back, to make top bearers or stretchers of the usual kind, and screw the top to them from below. The fronts of the drawers are 4in. deep and have a sunk bevel 1in. wide.

FIG. 301.—MOULDING ON TOP OF LOWER PART.

The framing of the doors is 3¼in. wide, except the top rail, which is about 1¼in. more at the ends. The easiest way to form this shaped rail is to make the rail as if it were to be straight, and then glue the corner pieces on as shown by Fig. 302, where the joint is indicated by the dotted line. It will also probably be

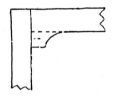

FIG. 302.—TOP CORNER OF DRAWER.

found easier to work the mouldings from the solid, at any rate at the curved parts, than to plant them on. The bevelling round the door-panels is 1¼in. wide, slightly sunk like the drawers.

On the ends the lower edge of the top rail is on a line with the bottom of

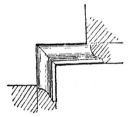

FIG. 303.—MOULDING ON FRAME.

the rail above the doors. The middle rail separating the two panels and the bottom one are the same width as the bottom one in front. The bottom rails come to within 1in. of the floor, but

as this is only sufficient to allow dust to accumulate underneath and not enough to allow it to be removed, the rails should either come to the floor or be raised considerably higher. All framing in this portion as well as in the others has a moulding round it, as shown in Fig. 303.

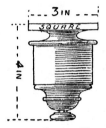

FIG. 304.—TURNED KNOB.

The middle portion of the cabinet is somewhat more complicated in construction, as it virtually consists of two parts, which, however, are fastened together to form one. It is 1ft. 9in. high, 1ft. 3in. from back to front on the cupboard or lower part, and 1ft. 7½in. on the moulding above the frieze.

The uprights of the framing at the front and ends, the drawing shows, are of the same substance as the corresponding parts of the lower portion. The top rail of the framing and ends forms the frieze and principal portion of the cornice on which the mouldings are fastened. It is 3¼in. wide. Below the frieze is a turned knob (Fig. 304), which is added on when making up the cabinet.

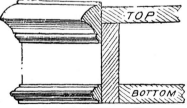

FIG. 305—TOP OF MIDDLE PART.

The mouldings are fastened on to the face of the frieze, as represented by the section in Fig. 305. The upper one is a little above the level of the top of this middle part, so that the top portion fits within it. The lower moulding covers the nails which fasten in the bottom of the cornice. This bottom, it should be noted, does not run right to the back, but only as far as the front framing in which the three doors are placed. By this means the utmost possible height is gained for the cupboard, and no space is wasted.

The widths of the top and bottom rails of the front framing which hold the doors are 2in. and 1½in. respectively. The stiles between the doors and the door-framings are all about 2in. wide. The panels have a sunk bevel 1in. wide. The doors are hinged on in the usual way, but it is a common thing to find the small doors of similar cabinets merely hung on wires driven through the top and bottom framing. This may seem a crude way of working, but it is effectual, and much less

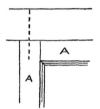

FIG. 306.—DOOR HINGED WITH PINS.
(AA, Door-frame.)

troublesome than fixing either "butt" or "centre" hinges. In case this simple form of hanging doors may not be understood, the accompanying diagram (Fig. 306), in which the wire pins are shown by the dotted line, will make the method clear. The principle is that of the ordinary centre-hinge, which is merely an improved form of the old arrangement.

With regard to the interior fittings behind the doors. The simplest arrangement is to have a shelf running the whole length of the cupboard. It is not then necessary to make a central door, as anything placed within the space thus formed can easily be got at through one or other of the end doors. The cupboard may be divided into three by partitioning off,

so that each door gives access to a separate compartment.

The construction of the third and

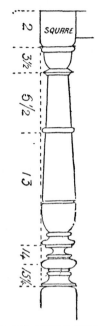

FIG. 307.—TURNED COLUMN.

top portion of the cabinet is very simple, and may be compared to the back of a modern sideboard. The

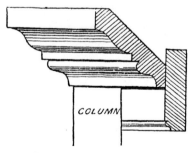

FIG. 308.—TOP MOULDING, ETC.

end-framing needs no remark, as its construction can be gathered from Figs. 299 and 300. There is no bottom attached to this part, as the top of the middle answers the purpose. The back has two panels. The shelf cutting across the panels is 6in. wide and about 7½in. above the bottom.

The columns are of 2in. stuff, and the pattern of the turning is shown in Fig. 307 on a sufficiently large scale, with measurements from the top to enable it to be set out full-size for turning. The frieze is 1¾in. wide below the cornice, with a moulded lower edge, and the moulding on top 3¼in. deep, as shown in Fig. 308. The top itself is divided into four panels running from back to front.

The thickness of wood for the various parts is not a matter of great importance. Speaking generally, if 1½in. "stuff" is used for the thicker portions of the work, such as the framing, it will be sufficiently stout. For the panels ½in. stuff will do very well.

Care should be taken in selecting the brass-work, in order that it may be in keeping with the style of the cabinet.

There are many ways in which the cabinet can be finished. If an "antique" colour is desired a stain must be used; any good brown stain will do for the purpose. But it will look very well if unstained, and finished by merely rubbing it with raw linseed oil; this gives a dull surface, with a very pleasant tone of colour.

A FANCY CABINET.

THE cabinet as shown in Fig. 309 is intended to hang against the wall, and consists of three fair-sized and several shelves on which other ornaments can be placed.

Fig. 309 is drawn to scale 1in. to

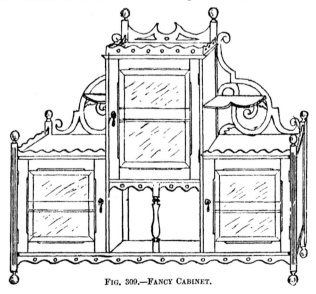

FIG. 309.—FANCY CABINET.

cupboards with plate glass doors, bevelled or not as may be preferred, 1ft.; the length being 3ft. between the uprights, and 7in. back to front. The side cupboards are 12in. high, the middle one 15in., all of equal width, so that, in marking out, the bottom board is divided into three equal spaces,

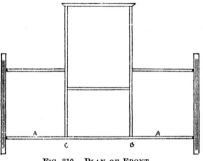

FIG. 310.—PLAN OF FRONT.

making allowance for the divisions between the cupboards.

Figs. 310 and 311 show plans of the front and back respectively. The work is as straightforward as possible, as anyone who cares to make the article will find out, all that is required being plenty of patience. For the back (Fig. 311) cut out four uprights from a 1in. board, 1in. square, two of them 2ft. 3in. long, and two 1ft. 6in. The thick lines shown on the uprights represent the sides of the end and middle cupboards; the horizontal thick lines represent the tops and bottoms of the same, all of which are of ½in. wood. The backs of the cupboards as shown are let into grooves cut in the uprights, and as the grooves must not extend the whole length of the uprights, they are best cut out by means of a cutting-gauge: this is practically a marking-gauge fitted with a long tooth. The groove is cleared out with a ⅜in. chisel, which will determine the width of the grooves, and a depth of ¼in. is sufficient. The backs for the side cupboards are 11in. high, 11½in. wide, the middle one being 14in. high and 11½in. wide. Glue the backs into the grooves, put the whole thing into clamps, and leave it for a time to set hard.

In the meantime the other parts can be got ready. Fig. 310 is a plan of the front. The bottom board AA is 3ft. long, 7in. wide, of ½in. stuff, the edges being either plain or fancy. Fig. 312 is a plan of the bottom board showing the position of the uprights, sides, divisions, and blocks. In each of the corner uprights a notch is cut into which the board is fixed, and the tops of the side cupboards are let in in a similar manner. The long uprights do not extend below the bottom board, but rest on and are fixed to it. Having then prepared the bottom board, mark out the positions for the divisions B and C, which are 2ft. high and let into the long uprights. The sides D and E are 12in. high, all the same width as the board A, and are let into grooves cut in both front and back corner uprights. It will thus be seen that the sides fit against the bottom, not on to it, for when using soft woods such as deal the cross-grain ends should be avoided as much as possible.

In Fig. 311 the thick black lines represent the tops, sides, and bottoms of the cupboards. Take the front uprights. The fronts and one side of each are ribbed with a bead-router by way of finish, and a shallow groove is cut in

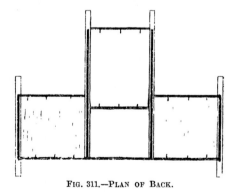

FIG. 311.—PLAN OF BACK.

the backs in which to let in the sides D and E (Fig. 312), and they are notched out to take the bottom board and the tops of the cupboards (*see* Fig. 313), which shows the back of an upright.

Next cut out the tops of the side cupboards and the bottom of the middle one. Fix the latter to the sides with screws, and then proceed to fix the

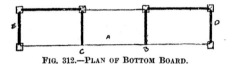

FIG. 312.—PLAN OF BOTTOM BOARD.

sides to the long uprights, gluing them into their respective grooves and securing them further with thin screws from behind. The tops of the side cupboards are fixed with screws to the sides of the middle one, and are also screwed on to the back and the sides D and E, all front edges being flush. The bottom board is next screwed on as shown, and the front uprights are fixed last of all.

The whole framework is very simple, but wants carefully fitting. Extend the grooves of the long uprights on the inner sides to the top, to take the fancy back above the centre cupboard. The top of the centre cupboard is a plain piece of wood with the edges chamfered,

FIG. 313.—BACK OF UPRIGHT.

as shown in Fig. 310, and is fixed with screws or wire nails. Use screws as much as possible for the main framework, as they will not be noticed.

The doors are made of ¾in. wood 1½in. wide, the rails being mortised

into the uprights or stiles, and a rebate cut on the inside for the glass. Cut the rebate as deep as possible, and when the glass is in secure it by a small beading on the inside. Each door is hung by a pair of 1½in. brass hinges; the machine-made ones are the best, being very neat in appearance. The side doors are hung on the uprights, the centre one either right or left hand, as may be preferred. Small turn buckles with fancy drop handles make a neat fastening. A small fillet can be run round the inside of the cupboards for the doors to shut against; but each cupboard has a shelf resting on small fillets screwed to the sides,

FIG. 314.—FANCY BACK.

and these will serve as stops for the doors if preferred. Speaking of the shelves, it is better for them to rest loosely on the fillets, so that they can be taken out if not wanted.

Above the side cupboards is a fancy back (Fig. 314). In order to cut the two pieces out, as well as the scalloped edging along the fronts and sides, it is almost necessary to have a fretsaw machine, or at least a fretsaw bow. In the event of possessing neither, the best plan is to cut out the outline with a frame saw, and leave the other part solid: some may prefer this. If a fret back is approved, a paper pattern must first be made of it in order to get both sides alike. The fixing is simple. A little strong glue and a few French wire pins such as cabinetmakers use, will make them as secure as necessary, as it is understood the cabinet must not be hung by them. Fig. 309 shows a shelf on each side fixed to the back and side of the cupboard. These are optional, as in the event of large ornaments being put on the top of the side cupboards they would be in the way.

The scalloped edging for all parts is 1½in. deep, of ¼in. or ⅜in. wood. The top edging is fixed on with glue and wire pins, the bottom also with glue, and in addition blocked at the back with small pieces of wood about 2in. long and 1in. square, placed at short intervals.

The small column beneath the centre cupboard is turned up out of any hard wood, being ⅜in. in the square. A groove is cut in the top of it to slip the scalloped edging in, so that it can be fixed in one piece (Fig. 315). The cabinet is hung by four strong looking-glass plates.

This cabinet can be made out of ordinary deal, with the exception of the knobs and the small pillar under the middle cupboard, and finished off with Jackson's Varnish Stain. If, however, black-and-gold is preferred, mix some vegetable black with warm

water, and apply it lightly with a soft brush or piece of sponge, being careful that it is put on equally all over; two coats will more likely ensure the black being regular. When dry give it a coat of size, and rub down all

FIG. 315.—PILLAR UNDER MIDDLE CUPBOARD

irregularities with fine glass-paper; then relieve parts of the work with gold paint, and finally give it one or two coats of good spirit varnish. If walnut or mahogany is used, it should be French-polished.

AN EASILY MADE DOG-KENNEL.

THE material used in making Fig. 316 is common match-boarding, which can be bought at a low price ready planed on one or both sides. For a kennel of moderate size, it must not be less than ⅜in. stuff. In many large towns American organ packing-cases are to be got at a nominal price, and these provide excellent material at a small cost, especially as the edges of the wood are generally matched. The size of the kennel will depend principally on that of the animal it is to house, and nothing need be said about it beyond the remark that it is better to have it too large than too small.

Cut all the pieces that are required to make the two long sides, being careful that they are sawn square and of the same length. The pieces are put together horizontally. The back and front pieces are vertical, and it will save trouble to make the width of the kennel such that it takes an exact number of planks, less the projecting tongue on one of the outer edges. There will then be no occasion to divide the width of any piece. The tongue can be planed off. The groove at the opposite edge of the back and front can be disregarded. It will be more convenient to fasten the sides on

N

to the ends than the reverse. The apex of the ends should form a right angle.

To stiffen the ends, a piece of wood may be nailed across all the portions forming them ; but this is hardly necessary, as the roofing binds them sufficiently, even though all the portions are of the same width, if a little discretion is used in driving in the nails. The sloping tops of the ends prevent the joints in these and the roof from being on the same line, as would be the case if the top were flat. To hold the ends together before the roofing is on, nails may be driven in at each joint. If the matching fits stiffly, this

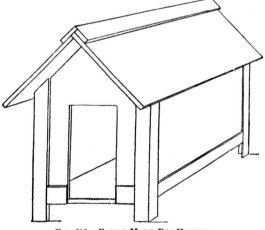

FIG. 316.—EASILY MADE DOG-KENNEL.

will not be necessary. The front end where the hole is will require a little more management. The opening can be more easily made with a square top than with a rounded one. It is formed by a piece or pieces of matching coming only part way from the roof, and, if necessary, cutting portions from the boards on each side of it. Just above the opening on the inside a piece of wood should be nailed across, to bind the pieces together. Screws may be used ; but French or wire nails driven through both thicknesses and bent over on the reverse side are quite as strong.

The roof must project at the sides,

back, and front, in order to carry off rain. A large projection in front prevents rain from drifting in through the opening. The tops of the sides may be bevelled to meet the slope of the roof ; but if the roof rests on their outer edges, this is not necessary. The tongue, which is on either the top or bottom edge of the sides, must be planed off. The roofing is laid with the pieces lengthwise. At the top the square edges (if the apex has been cut as directed) will meet exactly. Owing to the thickness of the wood, the roofing on one side will overhang more than the other. Whether this irregularity is observable or not will depend on the position the kennel occupies and its size. If it is objected to, the surplus wood can easily be sawn off. Along the top ridges two pieces of wood are nailed. It will be noted that they are so placed that the joint between them is not coincident with that of the roofing. They must be nailed each to the other and to the roofing, the nails projecting through this being clamped on the inside.

The bottom is formed by nailing pieces across from side to side. It will be better to fasten them on below than within the sides and ends.

Now, to give what may be considered the finishing touches, which not only improve the appearance of the kennel, but give it additional strength. The wood for the remainder of the work may be of the same thickness as that already used, or even thinner. Tongues and grooves will not be required. The latter may be left, as they can be hidden in every case, but the tongues must be planed off. Eight pieces will be required for the four corners—i.e., two for each. They must fit at the top against the roof, and must project below the bottom to serve as feet for the kennel. In each pair of pieces it will be better to have one piece wider by the thickness of the wood than the other : they give a finish to the raw corners of the kennel, besides rendering them watertight, which they might not otherwise be. To this end they must be so placed as to overlap like the capping of the roof. Pieces of similar width, say 2½in. or 3in., and of the same thickness as these, must be nailed on the sides and back to cover the edges of the bottom. Others may be put at the top ; but they are not necessary. The kennel may now be considered made, and only requires painting. About this, however, no remarks can be necessary.

A few hints may be given about the nailing. If the appearance of the heads of the nails is objected to, "cut nails," or brads, must be used in preference to French nails. They must be punched in to sink the heads, and the holes filled up with putty. Nails projecting through must be clamped by bending them or breaking them off on the inside. If the nails are driven in straight they will not hold so strongly as if slanting to and from each other, in which case it is almost impossible to pull apart work which has been so nailed together. Finally, let the length of the nails be at least double the thickness of the wood.

A HOME-MADE EASY CHAIR.

TO begin, obtain from a grocer a good-sized cask of light make, with wooden hoops. It must not be more than 2ft. high, the height from the floor to the row of hoops below the lid, or head, as it is called, being about 15in. Take out the pieces forming the head, and with a good, sharp saw cut down in the direction of the dotted lines (Fig. 317) to the first row of hoops, and then across, leaving an opening of rather more than a third the circumference of the cask. Care must be taken to secure the cut ends of the hoops which kept the head in by putting a few strong, short nails in at A and B ; this prevents the staves from opening, and makes the whole secure. It is now necessary to firmly nail a few pieces of wood or battens, like those used for supporting brackets, round the inside of the cask, in line with the first row of hoops at C, to form a support for the seat ; for these the pieces which composed the head will be found convenient and handy. Having done this and placed the seat in position, nail it here and there to the battens to prevent it from slipping. The cask is now of the shape shown at Fig. 318.

The next thing to be done is the padding, for which a mixture of horsehair and wool can be used. The seat must be well padded and slightly raised in a slope to the centre, and

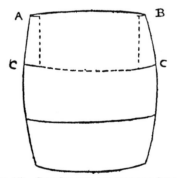

FIG. 317.—CASK SHOWING OPENING FOR SEAT.

then covered with a piece of wrappering, which must be firmly nailed all round. The padded cushion that surmounts the upper part must be made separately, also the upright piece that goes round the inside at the back.

N 2

Care must be taken that the padding is evenly distributed, and that no lumps or uneven surfaces are visible. If this part is "stabbed" here and there and buttons sewn on, it has a very good effect. They must then

the lower part may be covered with deep upholstery fringe, which may also hang from the arms. This also, with gimp, may be used as taste may direct, and forms a very handsome finish.

FIG. 318.—CASK WITH SEAT FITTED.

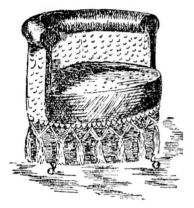

FIG. 319.—EASY CHAIR COMPLETE.

be very securely nailed on in their places.

Any material may be used for the covering—plushette is very suitable; and as the new-made chair has no legs,

If desired, some pieces of wood may be nailed on to the bottom, to which casters can be fastened, which makes it easier to move about. Fig. 319 gives the result when finished.

A COMBINATION CUPBOARD AND CHEST OF DRAWERS.

FIG. 320 gives a general idea of a useful piece of furniture designed to suit a children's nursery. Below is a chest of three long drawers, which is made quite independent of the cupboard, and is intended to contain children's clothes. The cupboard is made for their toys and for the nursery tea-things, etc., being fitted with a row of cup-hooks at the back.

The material used is good yellow deal. Give it two coats of boiled

linseed oil when finished, allowing each coat time to be absorbed and dry well, and then give it two coats of good, hard, oak varnish. Of course, choice of material is optional.

The chest, without the top, measures 3ft. 2in. long, 17½in. wide, and 2ft. 9in. high. The three drawers are of equal height in the front, viz., 8½in., and allowing for four rails (A, Fig. 321), each ¾in. thick, it leaves about 4½in. at the bottom: this is filled up with a plinth, or skirting, to keep the bottom drawer at a convenient height from the floor.

Fig. 321 shows clearly how the framework is put together, the front and back being made alike. On the left-hand side it will be seen how the sides are put on, together with the runners for the drawers. Wood ¾in. thick is used for the frame, and 3in. wide for both the rails and the uprights, the former being mortised into the latter, as shown; the top rails are dovetailed. It will be noticed that the uprights are not so long as the sides, though there is no particular reason for this. The front and back of the framework are made separately and screwed on to the sides, being flush at the front, but the sides must project ½in. beyond the back frame, to allow for the back. Underneath the bottom rails nail on a bottom; this keeps dust from getting into the drawers, and is best made of ⅜in. or ½in. match-board, on account of being tongued and grooved. The sides of the chest are 17½in. wide, which practically means two 9in. boards joined together; use ½in. stuff for these, and simply a glued joint—very little glue. With a true edge the join is hardly noticeable, and is very strong.

The drawer runners (marked B, Fig. 321) are fitted in between the rails, glued to the sides, and further strengthened by small blocks of wood glued underneath them. Take care that the blocks do not project so as to be in the way of the drawers. The rails will, when set, be very firm, and will require no other fixing.

Having screwed on the sides, the plinth, or skirting, is next fixed. This is made of ½in. wood, 4½in. wide, or at least wide enough for the top to come half-way up the bottom rail. To keep it in its place use a few French nails with their heads off; then, turning the frame upside down,

FIG. 320.—CUPBOARD AND CHEST OF DRAWERS.

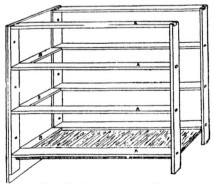

FIG 321 —FRAMEWORK OF CHEST.

block it on the inside, using plenty of glue.

The top of the frame is of ⅝in. wood, and is made large enough to project 1in. each way beyond the frame, the edges being chamfered or moulded with an ovolo plane. If this is not to

FIGS. 322 AND 323.—OVOLO MOULDINGS.

hand, a simple method is to cut a rebate, say ½in. wide and ¼in. deep, and round off the edges. Figs. 322 and 323 show a section of the ovolo moulding and the makeshift one respectively. Screw the top on to the frame from underneath; then fix on a back of thin match-boarding, nailing it to the back rails.

FIG. 324.—SECTION OF DRAWER FRONT WITH DOVETAILED SIDE.

The drawers are all of one size— 8½in. fronts, made of ¾in. wood. The fronts must first be cut and fitted. The sides and back are of ¾in. wood. The bottom, of ¼in. wood, is run into grooves cut in the sides and front, about ⅜in. up, and should slide in after the back is put on, being finally secured with a few wire brads. A thin

FIG. 325.—DRAWER FRONT SHOWING GROOVE INSIDE.

brace will be wanted to screw across the bottom boards to prevent them from bulging. Fig. 324 shows a section of the front with the side dovetailed into it; Fig. 325, the inside of the front and one side, showing groove for the bottom.

Guides must be glued on the runners on each side, so that the drawer shall not have too much play. If the drawers are not so large as the frame from back to front, fix two small stops on the front rails, for the front to shut against. This completes the lower half of the combination. Next comes the cupboard, which is made independently of the chest of drawers; but it is as well to fix the one to the other with a couple of screws, to prevent any shifting, and these can easily be taken out when required. The cupboard is 3ft. 10in. high, and the same width and depth as the chest of drawers, taking two 9in. boards for each side, of ½in. stuff, with a ¼in. bead run on the front edges. The top is nailed on to the sides, and the back, of match-board, is nailed on to the sides and top, the rough edges of the top being hidden by a moulding as shown; this moulding is 1½in. wide, and can be bought in lengths from any timber yard. On the front, above the doors, is a piece of 2in. square quartering, with a bead on the lower edge, and fixed with screws or nails from the top; and on the under-side of this, and 1in. from the front, is screwed a piece of 1in. square wood the whole length, which forms a stop for the doors to shut against.

The bottom, of ½in. stuff, is nailed on between the sides with good wire nails with the heads taken off, so that they can be punched in out of sight. The bottom is 1in. less, back to front, than the sides, to allow for the doors, which shut against it. Within the cupboard, a stretcher, 3in. wide, is screwed across the back, about midway, which serves to strengthen it, and in which brass cup-hooks are screwed at convenient distances for the cups. The cupboard is also fitted with two shelves, resting on fillets screwed to the sides; leave the shelves loose, so that they can be readily taken out to be washed at any time.

Finally come the doors, the stiles and rails of which are of equal width, viz., 2in., and are of ¾in. stuff. A groove is cut on the inside edge of each, ¼in. wide, for the panels. The stiles and rails are mortised together; the tenons do not appear through, but should be made as long as possible and fit well. The panels are of ⅜in. wood, bevelled down at the sides and ends to fit into the grooves, after which the doors are put together, the tenons well glued, and the whole thing clamped

together until dry. The inner stiles of the doors must be rebated, the one to shut against the other, as shown in the section Fig. 326. Each door is hung with three 2½in. brass butt hinges ; the left-hand one has two small spring bolts, top and bottom, to fasten it, the other being fitted with a 3½in. double-handed iron cupboard-lock and a small 1⅜in. brass screw-knob. The drawers have a pair of fancy brass drop handles on each. These, of course, are a little extravagant, but they give a finished appearance and are far preferable to the usual knobs.

In place of the cupboard, the amateur might prefer having a bookcase, which only means a slight difference in the pattern of the doors, a few more shelves, and perhaps the cupboard itself need not be so deep. For a bookcase the doors must be glazed, and as many amateurs do not possess sash-moulding planes capable of being used for cabinet work, Fig. 327 shows a simple, but at the same time a very effectual, door, 3ft. by 1ft. 6in., suitable for the purpose, having stiles and top and bottom rails 1½in. wide, with a deep rebate for the glass, and a small bead run on the outside edge. The sash-bars should be ½in. wide on the front, with a ₁₆³in. bead

on both edges. The glass is held in by nailing thin strips of beading behind, so as to form a bead on the inside corresponding with the front. The top of the door, by way of variety, could be arched as shown. The small corner-pieces are glued on the frame, and the glass cut to their shape. For a bookcase, dark oak, mahogany, or walnut stain is preferable to varnishing, and it will be found that Jackson's Powder Stains, which simply require mixing with hot water, are very good if put on without sizing the wood, so that the stain soaks well in ; though more stain is used, it stands the wear so much better. On this rub in a little

linseed oil, which takes off the extreme dulness. This is preferable to varnish ; and by occasionally applying a little oil, as one would a furniture polish, the wood retains a fresh appearance.

A SMALL WARDROBE.

F IG. 328 shows the wardrobe completed.   In the lower part is a
drawer, and above it is a cupboard with
what is known as a glass door, the panel

any space, and also considerably re-
duces the amount of work.   Even if
not to be placed in a recess, the straight
sides will not look unsightly.

The size of the
wardrobe is a matter
of some importance,
and although one of
this construction
might be made 4ft.
wide, one of 3ft. will
be more generally
convenient; but the
maker may expand
the size according to
circumstances, as the
construction will be
the same, with one
slight alteration
which will be found
noted in due course.

The height may
be about 7ft. 6in.
if a very tall ward-
robe is required,
or about 6in. less if
one of fair propoi-
tions will do.   The
depth from back to
front may be about
1ft. 6in., which, for
a wardrobe of this
kind, will give a good
roomy cupboard.

Ash, American wal-
nut, and mahogany
are the chief kinds
of wood used for bed-
room furniture, but
if pine or American
white-wood is used
it can be painted.   If
the last-named is
chosen, there ought
to be no great diffi-
culty in getting it of
sufficient width   to

FIG. 328 —WARDROBE.

being a mirror.   There is no cornice or
moulding round the top, as its absence
allows of the wardrobe being fitted
close within a recess without wasting

avoid the necessity of jointing for the
ends.   Even if made of a superior wood,
the back, top, and bottom, as well as
the piece between the drawer and the

cupboard, may be of pine on account of its smaller cost. Although slighter material might suffice, the most suitable for all the main portions of the work is 1in. stuff, which, when planed up, loses about ⅛in.

Having determined exactly the dimensions of the wardrobe, first do whatever jointing may be necessary to get the required width for ends, top, and bottoms. Plain glued joints are sufficient if properly made. They must be allowed to become perfectly firm before the wood is further worked on. The small projections on the upper edges of the ends must be cut out separately, and then glued on ; this is much simpler than cutting away the front edges to form them.

The top and two bottom boards are a trifle less in width than the ends, in order to allow for the thickness of the back, which must lie within the ends. As their front edges will show, they must be faced with the same wood as the rest of the work. This facing may be of veneer, but it will be better to use something thicker. Generally at least ¼in. stuff is used, but it is better to have it thicker. It is merely glued on.

The boards, after the facing has become firm on them, must be fastened to the ends by mortise and tenon, care being taken not to cut the mortises through the ends. Another joint, and one which may be recommended in this case, is to dovetail the ends of the top, etc., and fit them into correspondingly shaped grooves in the ends. This joint, as it binds the ends together, is stronger than the other, which, however, is strong enough, under ordinary circumstances. The top board is 6in. from the top and the bottom board 4in. from the bottom end. The distance from this latter to the one above it determines the depth of the drawer, and may very suitably be about 1ft. 2in. It may here be said that while squareness at all the angles is desirable, it is, if possible, more important to have the two bottoms accurate than the others. If they are not, it will be almost impossible to make an easy-running drawer.

In order to further stiffen the work, blocks must be glued at intervals into the angles above the top and below the

bottom. At this stage the work will appear as in Fig. 329.

The two side-pieces in front are merely plain boards accurately fitted in between the top and bottom of the cupboard, to which they may be fastened by two or three nails through each. At the outer edges they should be glued to the ends. The appearance

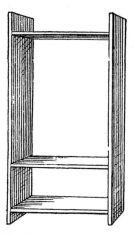

FIG. 329.—FRAMEWORK OF WARDROBE.

of panels is given to these pieces by gluing a small moulding as shown in Fig. 330. For a small wardrobe the plain boards as suggested do satisfactorily, but for larger sizes it will be better to make frames as for a door and fit panels within them.

The door may be made after the side-pieces are in, as it can then be fitted

FIG. 330 —MOULDING TO FORM PANELS.

exactly. Its width may be taken as 1ft. 6in. ; and as the construction is of the ordinary kind, nothing need be said about details. The glass fits within a rabbet, and is covered behind with thin boarding fastened to the back of the frame. The door is hung with butt hinges, which must be let into it, and not into the piece to which it is attached. Over the other edge of the door a narrow piece of moulding must be glued, so that when the door is shut it overlaps the side-piece. The best way to fix this mould-ing is shown in section in Fig. 331. If

FIG. 331.—MOULDING FOR DOOR.

a lock is used, it should be of the kind known as a "cut cupboard lock with bolt shooting to left."

The construction of the drawer calls for no special remark, as it is of the ordinary kind.

The only parts still to be attended to are the ornamental details at the top and bottom. The piece at the bottom is a very simple affair, being perfectly plain with the exception of the slight shaping at its lower edge; this can be easily managed by anyone who has a fret-machine or a bow-saw. The piece may be fastened with glue and a few screws through the bottom board, and with a couple of glued blocks at the ends.

The piece at the top is very similar, only for the sake of appearance it is moulded, the precise details not being important. As a suitable moulding may not be readily obtainable, a plain piece of wood slanting slightly forward, with a projecting piece on top (Fig. 332), will not look amiss.

The back should be either panelled or fitted with match-boarding.

Inside the cupboard a row of ward-robe hooks should be fastened. They are made specially for the purpose. If the back is too thin to hold them securely they must be fastened first to a rail, which can then easily be fixed.

FIG 332.—MOULDING FOR TOP.

Polishing or painting will complete the work, though the wardrobe will be improved by covering it inside with glazed lining.

A SMALL UMBRELLA-STAND.

A REFERENCE to Fig. 333 will give a fair idea of the appear-ance of the umbrella-stand. It projects such a slight distance from the wall that the space occupied is very trifling; while any amount of accommodation which may be required can easily be provided for by making the stand to a proportionate length. Whatever its size, the general plan of construction will remain the same. The example given measures 2ft. 6in.

It will hardly be worth while to use an expensive wood, and none is better than pine or American white-wood. Both these are cheap and easily worked. They may be painted, or, if preferred, can be stained and French-polished to resemble mahogany or walnut.

In thickness all the parts are of ¾in. stuff, which, when planed down, will not measure much over ½in. If pine is used, it should be free from knots. American white-wood never has any knots worth mentioning, being one of the cleanest of timbers in this respect. The two upright pieces are 2ft. 6in. long, and the top back piece the same length. The width of each piece is 3in.

These are fastened together so that the middle of the top rail is 2ft. from the bottom.

The halved joint is used, *i.e.*, each piece where it crosses the other is cut out to half its thickness, so that, when completed, there is a level surface over all. Two or three screws or even brads will make the joint firm. If the

FIG. 333 —UMBRELLA STAND.

edges are stop-chamfered, as shown, the appearance will be improved, but if that cannot be managed they may be just rounded except where the joint is, or even left square. This latter, however, will look crude, and the little extra expenditure of work will be well repaid by the improved appearance.

The piece which confines the umbrellas must be 4in. wide, and have a series of holes cut in it as represented. These openings are shown alternately large and small, as this gives a more pleasing appearance than having them all alike. It will be noticed that the front corners are round in the larger holes, and that in the others, which do for walking-sticks, all of them are so. Those who have either a fret-saw or a bow-saw will have no difficulty in shaping them ; but a more workman-like, because quicker, way will be to bore holes with a good-sized centre-bit, and then saw up to them from the back edge of the wood (Fig. 334), removing the surplus with a chisel. In the case of the smaller openings, four holes will have to be

made with the centre-bit, and the waste wood chiselled out. The edges of the openings must be rounded off both above and below and well cleaned up with glass-paper. The top edges of the front and ends of this piece must be chamfered or rounded to match the other portions and fastened to the back with screws, which must be driven in between each opening.

The bottom board is the same width as the piece last referred to, or a little less. In length it should be just enough, with the addition of some small pieces to be referred to immediately, to reach the outside edges of the uprights, to the fronts of which it is screwed. A small rail, say, about 2in. wide, and fitting closely between the uprights, is then screwed on behind it. At the fronts and ends, strips are nailed to form a rim. These may be about 1½in. wide, and must be fastened on the edges, and not on the top of the board. The two end-pieces must be of such a thickness that they will shoulder well up to the uprights, with whose outer edges their outsides must be flush. At the angles where the front and end pieces meet, they must be mitred.

To keep things placed in the stand upright, small strips of wood may be nailed across the bottom board if considered necessary. Those who wish to have a metal pan can, of course, have it made to fit, but there is no difficulty in doing without.

The stand, being made, only requires painting, or otherwise finishing, and fixing in place. To fix to the wall only a few nails are needed. If these are brass-headed, and driven at regular intervals, there will be no disfigurement, but rather the reverse. Another

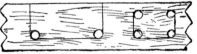

FIG. 334.—PIECE CONFINING UMBRELLAS.

good plan is to screw "glass plates" (so called because they are used to fasten looking-glasses to walls) on behind the stand, say one in the centre of the top rail and two to each upright, by which to fasten the stand to the wall.

By reference to Fig. 333, it will be seen that the bottom board does not rest on the floor, but is just above the skirting-board. The advantage of this is that it does not require the skirting-board to be cut away in order to allow the stand to fit close to the wall above it.

A PLAIN DRESSING-TABLE.

FIG. 335 gives the general appearance of the dressing-table when finished. It might be made, if desired, of the same wood as the small wardrobe described on p. 184, with which it would harmonise well.

The width of the top, or measurement from back to front, is 1ft. 8in.,

FIG. 335 —DRESSING-TABLE

the height from the floor 2ft. 6in., and the length 3ft. The jointing-up of the top is the first thing to attend to, so that the glue will have set firmly by the time this piece is wanted. It must be of 1in. stuff.

The legs are square in section, and are formed from 2in. stuff. The taper, which can easily be managed with the plane, may be made from each surface, but the appearance will be better if only the two inner faces are tapered, the outer ones being left straight. Naturally the top parts of the legs must be straight to an inch or so below the framing and drawer front.

The lower edges of the framing at the ends and back must be on the same level as the bottom of the front rail below the drawer, so the depth of this must be determined. About 4in. is a suitable width for the drawer-front, and as its bearer— the rail under it— is of 1in. stuff, the width of the framing at the ends and back is a little under 5in.

The top overhangs at the back as

well as at the ends and front, so the length of each piece of the framing must be set out accordingly. The exact dimensions will of course be got from the drawing, which must be prepared before the actual work is begun.

The framing is of ¾in. or 1in. stuff, and is fastened to the legs by dowelling, mortising, and tenoning, or simply sinking the ends and blocking them. The drawer-bearer need not be wider than the thickness of the legs, into which it should be tenoned. The rails connecting each pair of end legs, into which they are tenoned, are merely 1in. squares. The board which is laid on them and screwed from beneath is ¾in. stuff, and 9in. wide.

The drawer is of the usual construction and requires no special mention. The runners on which its sides rest and work may be glued to the framing. If it be feared that glue alone will not hold them sufficiently, they may be tenoned at back and front, or fastened there by a nail driven in slantingly. On top of these runners guides must be glued to prevent the drawer working sideways. The guides need not be more than ½in. thick. They must, of course, be quite straight from back to front, and flush with the surface of the leg against which the drawer works.

After this has been done, the top may be fastened on. This is managed by either gluing or screwing. The best plan, however, is to use screws, driven in slantingly, their heads being sunk. To allow for shrinkage of the top, the holes in the framing must be sufficiently large for the necks, or plain parts, of the screws to fit loosely within them.

A suitable size for the glass frame is 2ft. by 1ft. 6in. The material of which it is made is 1in. stuff, with a facing about 1½in. wide. In this case the outer edges will have to be veneered. The alternative of cutting the parts out of the solid and rabbeting them may be adopted, if preferred, and will be better if the table is of pine.

The arms supporting the bracket must be of 1in. stuff. Their shape is sufficiently shown in Fig. 335. The ordinary fretsaw frame and blade will not be sufficiently strong; but they can easily be cut with a bow-saw. They are fastened to the top by screws. The small shelves and supporting brackets are of ¾in. stuff, though there can be no objection to their being thicker. A dowel is let into the top and bottom of each bracket and fitted to corresponding holes in the shelves and table-top. The shelves themselves and the brackets are fastened to the arms by screws. To support the glass and allow it to swing, a pair of "glass movements," procurable from any dealer in cabinet brass-work, should be used. A part of each is fitted into the arm, and the other into the glass frame from behind. The movements must be fitted rather above than below the centre of the frame, so that the glass may not be top-heavy.

If casters are used they may be either of the "square-socket" variety or "screw casters." With the former the legs fit into the sockets, so these and the legs must be adapted to each other. The trouble of a misfit may be avoided by determining beforehand the exact size of the socket. In the case of screw-casters they are simply screwed into the legs, and for the sake of finish as well as to avoid risk of splitting them, caster-rims are generally used with them. The rims are to be had square, and are forced on at the extreme ends of the legs, so that the same precautions as are recommended for sockets apply equally to them.

A QUAINT WOOD CHIMNEY-PIECE AND OVER-MANTEL.

THE wood used is yellow pine, which in quality is one grade higher than best deal : it has the advantage of being free from knots and easily worked, and furthermore is inexpensive. This is required in three thicknesses, viz., $\frac{1}{2}$in., $\frac{3}{4}$in., and 1in. When finished, the wood is flatted with white lead and turps, the first coat being known as "priming," after

FIG. 336.—CHIMNEY-PIECE AND OVER-MANTEL.

which all "punch-holes" and—what may possibly occur—bad joints are stopped, and then the whole thing can be finished in any colour or tint that is preferred. Dead white is a good finish, and in applying the second and third coat of flatting, a little gold size should be added to harden it. Should it at any time become dirty, it can easily be made to look as fresh and new as ever by rubbing it down with fine glass-paper and giving it a coat of the flatting. This, of course, applies equally as well to any other colour that may be chosen.

be in fair proportion. It is necessary to mention this as the work should naturally suit the room for which it is intended. A high or a low-pitched room will make, or should make, a difference in the height; but the measurements given may be taken as a fair average. This piece of furniture is not intended for a large room, as it would be altogether out of character. It is suitable for a study, smoking-room, or drawing-room, not larger than 16ft. square, and for such a room has a very cosy and delightful appearance.

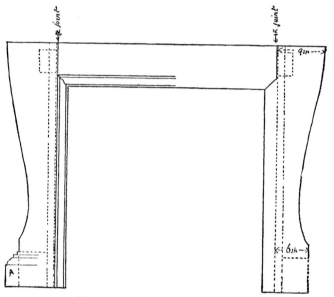

FIG. 337.—JAMBS AND CHIMNEY-BREAST.

With regard to measurements: the opening for the stove is generally a standard size, viz., 36in. by 36in., it being understood that the amateur who makes this chimney-piece and over-mantel intends it as a substitute for an ordinary one, which perhaps is not so sightly as it might be. It will therefore be necessary to conform to the usual size opening; with regard to the measurements of the rest of the work, however, they need not be followed accurately, but should always

The uprights on either side of the fireplace are the "jambs," the horizontal piece above the fireplace is the "chimney-breast," then comes the mantel-shelf, and finally the over-mantel.

Fig. 337 shows the jambs and breast of the chimney-piece. The opening given is 36in. square: this takes a Register grate 38in. by 38in., which is the regular size. The jambs will therefore overlap the stove 1in. on each side and the chimney-breast 2in.

The diagram shows the jambs 9in. wide at the top, not including the moulding, and 6in. at the bottom. The chimney-breast is 6in. wide, not including the moulding, and all is of 1in. pine ; the height of the jambs is 3ft. 8in. The left-hand side of the drawing (Fig. 337) shows how the moulding is worked

FIGS. 338 AND 339.—MOULDINGS.

on the jambs and breast, and for this an extra 2in. must be allowed in the width of each when cutting them out. The jambs and chimney-breast are mortised together, the tenons being shown by the dotted lines. The moulding is run on the whole length of each, and is afterwards cut off at a mitre angle, as also shown. The right-hand side of the drawing shows the jamb and breast fitted together without a moulding, but in this there is a mitre as shown, and an independent moulding can afterwards be fixed on. Figs. 338 and 339

be screwed across the opening to prevent the joints from getting strained.

The vertical dotted lines show the positions of the front uprights of the jambs ; these are also of 1in. stuff, and are cut out the same shape as the other uprights, and screwed to the jambs themselves from behind. As the front edges of the uprights come flush with the shelf, the width will depend upon the width of the shelf. The jambs at the top are, as shown, 9in. : the uprights are therefore the same size as the jambs from the outside dotted lines.

The jambs are finished off at the bottom with a moulded plinth, or skirting (marked A in Fig. 337) which is built up of two thicknesses of wood. First there is a piece of ½in. stuff, 6in. wide, with a square edge ; this is bradded on the jambs with mitred joints. Next comes a piece of 1in. stuff, 5½in. wide, with a moulding worked on as shown : this, of course, is prepared in one piece, then cut to proper lengths with mitred corners, and also bradded on, with a little good but thin glue in addition. Care must be taken in cutting the mitres so that

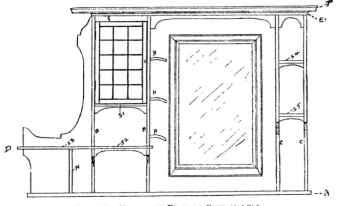

FIG. 340.—ELEVATION PLAN OF OVER-MANTEL.

show sections of suitable mouldings, the planes for either pattern in 2in. costing about 6s. each ; but similar mouldings can be bought in lengths at most timber merchants.

Having fitted the jambs and chimney-breast together, a temporary stay must

the joints fit nicely when the effect of the heavy moulding is very good. The corner brackets, two on each side, are screwed on from the back through the front uprights, and fastened by means of French nails with their heads off ; by punching the latter well in, the holes are easily stopped with putty so as not to be noticeable.

The mantel-shelf is of 1in. stuff and 5ft. long. The width is 9in., but in addition it should overlap at the back 1½in. for fixing in the wall; but if it is undesirable to touch the brickwork, then leave it flush with the back, and use strong, brass glass-plates for fixing. The left-hand corner of the mantel-shelf is rounded off in character with the brackets underneath, the right-hand corner being left square. This, so far, completes the mantel-piece, which is made independently of the over-mantel, although it is better to construct the latter also before fixing the mantel-piece.

Taking the mantel-shelf (A, Fig. 340) as being 9in. wide, it is better to make the over-mantel rather less from back to front—say 7in. at the widest part—and the height about 3ft., or in proportion to the height of the room for which it is intended. The four principal uprights must be cut out first, viz., BB, CC (Fig. 340), these are shown clearly in Figs. 341 and 342 ; the measurements being given in the drawings, it is not necessary to repeat them. These uprights are of ¾in. wood. G¹ is a single groove on the inside only, to take the bottom of the cupboard (all shelves being of ½in. wood) ; G² is a groove on each side to take the shelf D, Fig. 340. In Fig. 342 G³ and G⁴ are grooves on the inside only. On the outer side of the right-hand upright (B) cut three grooves at equal distances to take the small shelves (H). Now cut out the top board E (Fig. 340), 3ft. 10in. long and 7in. wide, of ¾in. stuff, and proceed to fix it on the uprights with good long screws through the top, keeping the uprights BB 11in., and CC 5in. apart. At the same time nail a temporary piece across the bottom representing the mantel-shelf, which can be removed when the over-mantel is finished. To the skeleton framework thus made the remaining parts can easily be added.

It is always preferable that the uprights of the over-mantel should fit close against the wall. In order to do this, the wooden back must not extend beyond it, but rather be fitted in between the uprights ; a neat fillet of ½in. (or ¾in. if ½in. wood is used for the back) must therefore be fixed on them. When putting in the back between BB and CC (Fig. 340), let the grain of the wood be perpendicular, so that each part may consist of one piece only. If, however, the amateur has not wood 11in. wide, it is easy enough to joint up two pieces to the required width, a glued joint being strong enough.

The back between the inner uprights B and C is about 2ft. 2in. wide, with an opening for the glass 26in. by 17in. (see Fig. 343). It is made of ½in. stuff, with glued joints at J. The moulding

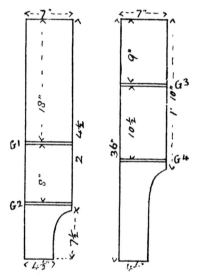

FIGS. 341 AND 342.—PRINCIPAL UPRIGHTS FOR OVER-MANTEL

forming the frame is fixed on the face, overlapping the opening ¼in. on all sides ; this forms the rebate. The dotted lines in Fig. 343 represent a back to keep the glass in position, and which in addition strengthens the back itself. The moulding should be 1½in. or 1¾in. wide, and should correspond in pattern with that round the top board. A bevelled mirror of the size required (26in. by 17in.) costs 10s. The back is now fixed in between the uprights, the fillets forming a shoulder against which to secure it. Now for the shelves. The measurements of

O

these have been given, except those of shelf 3, which is 15in. long ; this is supported by the upright N (Fig. 340), which is in a line with the upright

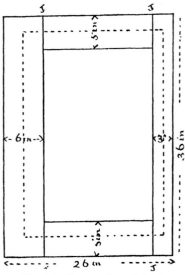

FIG. 343.—BACK OF OVER-MANTEL.

on the joint of the mantel-piece (Fig. 336). The left-hand wing of the back is screwed to shelf 3, as well as to the fillet fixed on the upright.

The door of the small cupboard is the most delicate piece of the whole thing, and calls for very careful work. It is 17in. high and 11in. wide, the frame 1in. wide and ¾in. thick. The door is divided into fifteen squares, and the sash-bars must be as thin as possible (as shown in section in Fig. 344, which is drawn large to clearly illustrate the fitting of the cross-bars). The horizontal bars being mortised into the frame of the door, the upright ones are then fitted in separately, being cut so that they will fit (as shown in Fig. 344), the rebate being on the inside of the door. Each upright sash-bar must be glued in, the work being too fine for nailing.

The squares of glass can be fixed in either by the ordinary method of glazing, using as little putty as possible (of course, after the wood has

had at least one coat of paint), or with fine beading ; but the former method is the better. The door is hung with a pair of 2in. brass hinges, with polished knuckles, and secured either with a brass turn having a fancy handle or with a small brass lock ; but in either case a fancy handle gives a finish to the work. A thin fillet must be fixed on the inside of the cupboard for the door to shut against.

Above the door is a piece of wood (F, Fig. 340) 1in. wide, fitted between the uprights and fixed with small blocks glued to it and the top of the cupboard. The same method of fixing is used for the scalloped pieces below the shelves and top-board. Finally, a second or additional top with a moulded edge (P, Fig. 340) is required : this is 4ft. 4in. long and about 8½in. wide. The moulding is 1¼in. deep, and as it is better not to cut this on the top itself a thin top will do, the moulding being fixed to it with mitred corners.

As the over-mantel is made independent of the mantel-piece, it must be provided with four glass-plates to hold it firm. The cupboard, too, may require a shelf or two, but these are mere matters of detail. The work must have a priming of white lead and turps, after which all holes and bad fits must be stopped with putty.

FIG. 344.—SECTION OF SASH-BARS.

Whatever colour is decided upon, the work must have at least two coats, sand-papering each coat down when dry, except the final.

A DRAWING-ROOM SECRETAIRE.

FIG. 345 gives a general idea of the secretaire, which is 4ft. 9in. in extreme height, or 4ft. 6in without the feet; 2ft. 6in. wide, and 9in. deep. The feet are 18in. by 3in. high, and the whole thing is run on small casters, for the convenience of moving it easily with the sides, the under-shelf, drawers, and partition being set back to allow for this. The back of the secretaire can be finished in various ways, although the one shown has only a plain back, enamelled, and draped with a sort of art silk in pleats drawn

FIG. 345.—SECRETAIRE

to any part of the room. The lower part, as far as the front is concerned, is panelled, as shown. Then comes the writing-desk, which is hinged so that when not in use it folds up, and fastens with a spring catch to the under-side of the book-shelf, and flush tightly, in which photographs can be stuck.

To begin. Cut the feet out of 1¼in. stuff, mortise the sides into them, and run a stretcher from end to end to make the secretaire rigid (Fig. 346): this also helps to support the bottom

o 2

board. The wood for the sides and shelves must be not less than ⅝in. thick when planed. When buying the wood get it machine-planed : it is then certain to be true, costs very little more, and only requires the smoothing-plane to finish it. The

FIG. 346.—FEET AND STRETCHER

edges of the sides and shelves are run with a beading, as shown in Fig. 347, by means of the bead-router.

With the exception of the bottom shelf, it is better to let the others into shallow grooves cut in the sides. Shelf A (Fig. 345) is 1½in. from the top. Allow 9in. between A and B, 6½in. between B and C, and 12in. between C and D. The desk is hinged to shelf D, and when closed shuts under shelf B ; so, according to the measurements given, the desk is 19½in. deep. Shelf C must be set back ¾in. to admit of the desk shutting flush with the sides ; and all shelves, including the bottom, must be set in ½in. at the back, in order that the back itself, which is ½in. thick, may fit flush with the sides. The back should be joined up (glued joints being sufficient providing the edges are true), or, to simplify matters, tongued and grooved boards can be used, as they are covered with the drapery,

FIG. 347.—EDGES WITH BEADING.

and consequently not noticeable. As the top edge of the back is cross-grained, put a small false beading round by way of finish.

The small columns between shelves B and C are turned out of hard wood, with square top and base, the top part having a cut through, as shown in Fig. 348 (which gives proper measurements throughout), so as to enable the edging E (Fig. 345) to be in one piece. Three

columns are required, one being sawn in halves to provide the end ones. As the grooves in the top part of the columns are ¼in. wide, the edging E must naturally be the same in thickness, but will require no fixing : the columns themselves being fastened top and bottom with screws through the shelves, will be sufficient, and they are fixed flush with the shelf C.

As regards the partitions and drawers, there is hardly any need to go into details ; the object is to make them as light as possible, though at the same time strong enough to be of use. The shelf D comes flush with the sides at the front, for to this the desk is hinged. The next part to make is the panelling below the desk. This is a fixture, consisting of a frame 3in. wide, and a centre stile of the same width, the inner edges being chamfered, and also grooved for the panels. The framework is of ⅝in. stuff, with a ¼in.

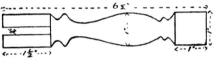

FIG. 348.—COLUMN PREPARED.

groove ; the panels are of ⅜in. stuff, bevelled down (Fig. 345) to ¼in., making the bevel 2in. wide. There is no occasion to give further measurement. The panelling must be fixed in its place from the inside before putting in the back.

Finally comes the desk, which perhaps requires the most care in making. In Fig. 349 a section of it is shown to give readers some idea how it is made. It is of the same thickness as the panelling below, the framework being also 3in. wide, with edges chamfered on the under-side, which is the front of it when closed, to correspond with the panelling. The grooves to hold the centre are rather wider, as the centre should be as strong as possible so as

FIG. 349.—SECTION OF DESK.

not to spring at all when in use as a writing-desk. The centre, it will be seen in Fig. 349, is also bevelled on the under or outside to match the panelling, the top or inside being left quite flat ; it must fit just below the surface of the framework, and on it should be fixed, with thin glue, a piece of green or any other coloured cloth.

The desk is hung with three polished brass hinges 2in. long, and supported by brass quadrants (Fig. 345) fixed flush on the sides. A small brass spring catch or small brass lock keeps it secure when shut up, and the secretaire is complete with the exception of finishing. First, then, give it a coat of white lead and turps, with a little boiled oil mixed in : this will be absorbed a good deal by the wood, and when dry must be well rubbed down with fine glass-paper, any imperfections or brad-holes being stopped with putty. Then give it a second coat, taking care to rub this down with glass-paper when thoroughly dry ; it will leave a nice smooth surface ready for whatever colour, either in paint or enamel, it is intended to use. To give an enamelled appearance, a good plan is to make up some white paint, strain it carefully through some fine gauze, then add some good copal or oak varnish. This will give an ivory-white colour and a bright surface, and one that not only looks well but will stand well ; further, should it get finger-marked, it is easily cleaned, and thus retains its fresh appearance. The desk cloth must not be put on until the painting is done.

HOW TO MAKE A MORTISE AND TENON JOINT.

FINAL INSTRUCTIONS.

THE method of properly cutting a mortise and tenon joint is so important that we cannot do better than conclude our little book with final directions on the subject. For the purpose of illustration two pieces of yellow pine, 1in. thick and 3in. wide are taken. The length may be 1ft.

Begin by planing the wood to an even surface. True the front side first, using the square to ascertain its correctness. This front side is of great assistance in putting together all varieties of framing and general work of every description. It is the point from which all the work starts, and if a mark—an ordinary cross, or whatever mark suggests itself—is put on this front side, it will act as a guide throughout the piece of work. Next true the remaining edges and surfaces.

With a square mark off from the end of the rail, on both sides, 3½in., shown by

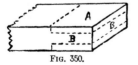

FIG. 350.

the dotted line A in Fig. 350. This gives the length of the tenon, but it will be noticed that the tenon is ½in. longer than the width of the wood · this is a safeguard for a well-fitting joint.

Draw the dotted lines B (Fig. 350), and mark with a gauge set to one-third the thickness of the stuff ; then with a tenon saw cut down the dotted lines A and B. This will give the tenon as shown in (Fig. 351).

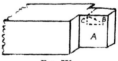

FIG. 351.

Then mark the dotted lines A and B (Fig. 351). First cut down the line A, and then cut diagonally the line B ; a saw-cut through the dotted line at the base of the tenon will detach two wedges, shaped as in Fig. 352. These wedges are the exact width

FIG. 352

of the mortise, and are of great service in wedging up.

The tenon, after the lines B and C have been cut, appears completed as in Fig. 353.

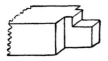

FIG. 353

It will be observed (Fig. 351, C) that at the bottom of the tenon ¼in. is left to form what is known as the "hauching," which when, the tenon is fitted into the mortise, is "stumped" in to fit, in other words a small trench is made to receive it. The object of this is to strengthen the joint. Any little irregularities may be trimmed off with a broad chisel.

Take the other piece of wood, and mark with a gauge an opening exactly the size of the tenon just made (Fig. 354, A), leaving 1½in. on the end of the stile

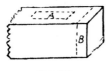

FIG. 354

(Fig. 354, B), to prevent splitting when wedging the tenon in. Then with a mortise chisel, the exact width of the mortise, proceed to cut out the necessary space, the bevel of the chisel being kept away from the edges of the mortise.

At each end of the mortise make a slight bevel into which the wedges may fit.

If the work has been accurately done, the tenon will fit into the mortise and appear as in Fig. 355.

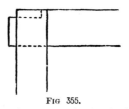

FIG. 355.

Glue the tenon, insert it in the mortise, and place in a cramp. Then glue the wedges (Fig. 352), and drive them in at the bevel on the mortise. The joint is then quite firm. After the joint has become set and dry, the ends of the tenon, stile, and wedges may be sawn off, worked to a level surface with a smoothing plane, and afterwards carefully sand-papered.

"SETTING OUT."

The greatest attention must be paid by the amateur to what is technically known as "Setting out." Both the tools and the wood may be good, but only indifferent results will be obtained if an accurate setting out has not been made in the first place. Each part of any piece of work must be set out in full on a spare board, or on the wood it is to be made out of, and the following points borne in mind.

(1) Allow sufficient length in the stiles, say 2in. over and above their actual demensions, so as to provide against the mortise splitting in the process of wedging up.

(2) Allow about ½in. over and above the actual measurement of all tenons.

(3) Allow for the hauching in the stiles.

(4) Allow in the stiles and rails in the event of a plough groove being required for the insertion of panels.

(5) Allow, in the case of framing, &c., where two stiles form a corner, a width equal to the thickness of the stuff.

(6) Allow for the kerf or passage of the saw.

Attention to these points will save much time and prevent disappointment, whilst at the same time making a basis for really good work.

INDEX.